DATE DUE	JAN 1 3 2000		
DEC 1 4 2004			

Also by Michael D. Lemonick

The Light at the Edge of the Universe:
Leading Cosmologists
on the Brink of a Scientific Revolution

OTHER WORLDS

The Search for Life in the Universe

* O *

Michael D. Lemonick

Simon & Schuster

SIMON & SCHUSTER
Rockefeller Center
1230 Avenue of the Americas
New York, NY 10020

Copyright © 1998 by Michael Lemonick

Designed by Sam Potts
Manufactured in the United States of America

1   3   5   7   9   10   8   6   4   2

Photo credits appear on page 259.

Library of Congress
Cataloging-in-Publication Data

Lemonick, Michael D., date.
Other worlds: the search for life in the universe/
Michael D. Lemonick
        p.   cm.
Includes bibliographical references and index.
        1. Life on other planets. I. Title.
√ QB54.L43   1998   97-49006   CIP
        999—dc21

ISBN 0-684-83294-1

111441

ACKNOWLEDGMENTS

This book would have been impossible to write if a very large number of friends, colleagues, acquaintances, and (at the time) perfect strangers hadn't been willing to share their time and energy so generously with me. My deepest appreciation goes to my literary agent, Cynthia Cannell, whose sharp insights and consummate professionalism were indispensable in turning a vague idea into a solid book proposal; whose encouragement and general hand-holding got me through the rough parts; and who never failed to treat me as though I were her most important client—an impression I'm convinced she gives to all of us. Martha Johnson was equally professional as her thoughtful and tireless assistant.

I'm also in debt to my editor, Robert Bender, whose acute eye for language and exposition kept me honest and who worked with me to make this book as good as it could possibly be. Thanks to Johanna Li, the mechanics of the editing process were so smooth that I barely noticed them.

My thanks go as well to the astronomers and other scientists who took the time to speak with me and to give me the insider's tour of their lives and work. First among them are Geoffrey Marcy and Paul Butler, who let me intrude on their terribly busy schedules over and over again. Among the others who made time for my endless questions were Roger Angel, Gustav Arrhenius, Peter Backus, Jeffrey Bada, David Black, Bill Borucki, Alan Boss, Bill Cochran, Mark Colavita, Kent Cullers, Frank Drake, John Dreher, Larry Esposito, George Gatewood, Everett Gibson, James Head, Gerald Joyce, James Kasting, David Latham, Hal Levison, Doug Lin, Jack Lissauer, Andrew Lyne, Buddy Martin, John Mather, Michel Mayor, Stanley Miller, Steve Mojzsis, Leslie Orgel, Tobias Owen, Mike Shao, Seth Shostak, Jill Tarter, Richard Terrile, Steve Vogt, Gordon Walker, Ed Weiler, George Wetherill, and Alex Wolszczan. Many of them endured a second round of imposition by agreeing to review parts of the manuscript in order to detect the inevitable errors I introduced in the writing. Those mistakes or misperceptions that have snuck into print despite their best efforts are entirely mine.

For logistical help in reaching these scientists and reaching the nearly inaccessible places where they sometimes work, I thank the support staffs at all the institutions and universities I visited in the course of conducting interviews. All were unfailingly helpful, but I want especially to thank Andy Perala of the Keck Observatory and Steve Maran of the American Astronomical Society for their efforts on my behalf. Philip Elmer-DeWitt, Chris Porterfield, Jim Kelly, and Walter Isaacson were extremely generous in giving me the time I needed to give this project proper attention.

I'm indebted to Heather Barros, Laurie Foor, Brad Hill, Ben Hohmuth, and Jeffrey Kluger for their enormously useful comments and conversations about the work in progress. They saved me from more than one dead end.

I thank my extended family—all the Lemonicks, Drutts, and Bernsteins, and their children and grandchildren—whose support is ever present. I thank my father, whose counsel, company, and love I treasure. And most of all I thank my immediate family, Eileen, Ben, and Hannah, who are always there for me, no matter what.

For Eileen,

whose love and support
make life on this world
far easier to manage

OTHER WORLDS

CONTENTS

INTRODUCTION

Paul Butler sits at his desk, utterly still. He's transfixed by the image displayed on the computer in front of him. The screen shows a field of dots. To the untrained eye, they seem utterly random, like flecks of black pepper sprinkled on a sheet of white paper. They look that way to Butler's exquisitely trained eye as well. He's usually very good at spotting patterns in data like this; the hidden shapes tend to jump right out at him. This time, however, if there is order underlying the apparent chaos, it's beyond him—and therefore probably beyond the capacity of any human being.

He has a software program, though, that can make sense of this mess, provided that there is any sense to be made. What Butler hopes—assuming that the last eight years of painstaking, frustrating, rarely appreciated work has not been a complete waste of time and that nature will cooperate, which is by no means guaranteed—is that the dots will fall along a curve. He doesn't know exactly what this curve will look like, but broadly speaking, it should snake from left to right, rising toward the top of the screen, then falling gracefully to the bottom, then rising again, over and over. It should resemble, in a very rough way, the reassuringly regular pulse of a healthily beating heart displayed on the cardiac monitor of a hospital coronary care unit. Most important of all, the curve should pass through just about every one of the dozens of spots on the screen, linking them together as though this were an incredibly difficult connect-the-dots game whose picture is impossible to pick out in advance.

If the curve exists, concealed in the data, his software will be able to find it and trace it for Butler's merely human eyes. He was here very late last night, looking at screen after screen of data for such a tracing, plugging new information into the computer, scanning the results, moving on to the next data set. His colleague and mentor Geoffrey Marcy—like Butler an astronomer with joint appointments at San Francisco State University and the University of California, Berkeley—was working alongside him until midnight. By two in the morning, Butler, too, had to quit. He got on his bike and pedaled back to his tiny apartment in the south Berkeley ghetto (his own description of the neighborhood).

Now, only six hours later, he is back at the task that has consumed him for the past two months. While Butler slept, the software was patiently searching for that elusive perfect fit, a curve that would slice neatly through most of the dots on the screen. It has tried fat curves and skinny curves, examining each one and discarding it when too many dots were left floating, unanchored.

The program finished its calculations a few minutes ago. Butler is looking at the result. There is the line, as lusciously curvaceous as the illustration in a textbook. Every dot on the screen is sitting right on the line or very close to it. No stragglers. This is precisely the pattern Butler has been aching to see. And now that he is looking at the squiggle that has haunted his dreams since he began this project eight years ago, all he can do is stare. The Berkeley campus is slowly coming to life on this winter Saturday morning, but Butler is completely unaware of it. He hears nothing, feels nothing, sees nothing but the glow of a cathode-ray tube bearing news that will forever change humanity's understanding of where it stands in relation to the cosmos.

The dots on Butler's computer screen mark observations of 70 Virginis, a star quite similar to the Sun in mass and age and temperature—pretty much a garden-variety G star, according to the arcane classification scheme astronomers use. The star 70 Vir lies in the constellation Virgo and is perhaps thirty-five light-years away from Earth—two hundred trillion miles, more or less.

The graph on Butler's computer screen represents 70 Vir's motion in relation to our own solar system, plotted over several years. If a dot appears at the top of the screen, that means the star is moving toward us. If a dot shows up at the bottom, it's moving away. The fact that every dot in Butler's plot falls on a perfect curve, though, means that 70 Vir is moving both toward *and* away from Earth—first the former, then the latter, in a rhythm that repeats itself every 117 days. 70 Vir is wobbling.

It's no surprise to discover that a star is moving. Every star we can see is moving as it orbits in an enormous, lazy circle around the pinwheel-shaped Milky Way galaxy. They're all like horses on a merry-go-round that measures one hundred thousand light-years, or about six hundred quadrillion miles, across. Unlike the horses on a carousel, though, the stars are not bolted in place. Each star orbits independently, following its own unique path through the heavens. It's true that the stars appear to be unmoving: The constellations, Virgo included, look precisely the same tonight as they did last night, and a hundred years ago, and a thousand. But that's really an illusion created by the fact that the stars are incredibly far away and that we're all moving in about the same direction at about the same speed.

Not precisely, though. Every star is constantly being affected by the combined gravity of hundreds of its neighbors as well as the gravity of huge clouds of interstellar gas. The stars are actually weaving in and out as they go, in a slow dance, essentially imperceptible due to the vast distances between them. Tens of thousands of years from now, the constellations that Chinese and Arab astronomers knew intimately—and that we in the Western European tradition still call by their Greek and Latin names—will have lost their shapes and taken on new ones.

For the very closest stars, astronomers can actually measure these gradual motions. The fastest-moving star on record, Barnard's Star, shifts its position

by about one two-hundredth the width of the full Moon each year. Astronomers have been making such delicate measurements for many
decades. But the path a given star traces across the sky should be smooth,
not bumpy. It should, that is, unless something is yanking on the star. Outside the solar system, the Sun's motion would be bumpy because the Sun
has planets orbiting around it. The planets are relatively puny; even Jupiter,
by far the biggest, is only one one-thousandth as massive as the Sun. Still,
their gravity is enough to have a small but distinct effect. From a light-year
or two away, the planets themselves would be invisible, but their presence—
especially that of Jupiter—could be inferred from the wobbling of the Sun.

Now Paul Butler is looking at 70 Vir. The computer is telling him that 70
Vir is wobbling. Of course, the computer's opinion is only as good as the
data it has to work with. Butler and Marcy are the ones who made the telescopic measurements of 70 Vir's position over the years. They designed and
built some of the instruments they used. Butler wrote much of the software
for his data analysis—thousands upon thousands of lines of arcane program-
ming language. If the computer is wrong, it isn't the computer's fault.

Butler and Marcy are extremely careful scientists, though. They've tried
to think of every conceivable way they might be fooling themselves, and
there are many. They know that if they can find planets circling stars beyond
the Sun, they'll earn tremendous respect from their colleagues, public ac-
claim, and—perhaps most important—a flow of research funding to keep
their work going. They know equally well that if they announce a discovery
that turns out to be wrong, they'll be cited in basic astronomy texts for the
next fifty years as examples of how not to do science.

So they've tried to be harder on themselves than any critic in this extraordinarily competitive field would ever think of being. They've second-guessed
themselves at every turn and are convinced they've eliminated every source
of self-deception that might exist. The dots are so beautifully studded along
the curve that Butler can see right away this is what scientists call a forty-
sigma detection. "It was roughly as unmistakable," Butler would say later, "as
an atomic bomb blast."

A forty-sigma detection means that the wobble they've measured in 70

Virginis's position is forty times as large as the uncertainty in the measuring technique. "Most cutting edge detections in science," Butler would explain, "are of the three-sigma variety. If the size of the signal you are looking for is only three times the size of your errors, the detection is only marginally believable." It's as if you had a speedometer that was accurate only to within 20 miles per hour. The speedometer might read 60 when you were actually going anywhere from 40 to 80. With a three-sigma speedometer, you'd be a fool to feel secure while zipping past a police car. With a forty-sigma speedometer, on the other hand, a reading of 60 would mean you're going between 58½ and 61½. You could be pretty sure of yourself telling a cop you were doing 60. If the astronomy police pulled Butler over right now, he'd be on safe ground. There appears to be no other plausible interpretation of what he is seeing: 70 Vir is not alone in space.

"My mind simply shut down," he will say a few months later when things have finally begun to return to normal. "Archimedes jumped out of the bathtub and ran naked through the street shouting 'eureka.' But I just sat there, dumbfounded. Nobody was around. I'd blink, and it was still there. I couldn't move. I could only turn my head."

After about an hour, the phone rings. His girlfriend, Nicole, wants to know what's up. He can barely speak a coherent sentence. After she hangs up, it occurs to him that he should let Geoff Marcy know what has happened. The two astronomers have gotten to be good friends, working together for almost nine years, but given the long hours they both put in, they're very respectful of each other's off-duty time. Butler hates to disturb Marcy at home. He does it anyway.

At his house in Berkeley, Marcy is heading out the door with his wife, Susan, when the phone rings. They're on their way to buy supplies for a New Year's Eve party the next night. He hears Butler say, in the most serious tone of voice Marcy has ever heard him use, "Geoff, get over here." Then silence. Marcy has to pry the information from him. "What is it, Paul? Have the computers been stolen? No? Is the building on fire?"

After several minutes of trying, the most he can get from Butler is that the news is good. "Come *now*" is all Butler can manage to say. So Geoff and Susan drive over to campus, burst into the office, and find Butler still looking at the screen. The data are still up on the computer. Marcy looks for approximately a second, then says, "Oh, my God."

∗ ◯ ∗

Kathie Thomas-Kaperta has a decision to make. She's supposed to present a paper at the annual Lunar and Planetary Science conference, where researchers will gather from all over the world to talk about and listen to and argue over the latest results on everything from newly discovered comets to lightning on Jupiter to the moons of Uranus. Thomas-Kaperta's paper has an innocuous enough title: "Microanalysis of Unique Fine-Grained Minerals Within the Martian Meteorite ALH84001." What it says is that Thomas-Kaperta, a geologist with the Lockheed-Martin Company's space science division, has used a powerful device called a transmission electron microscope to peer inside a slice of rock thinner than a piece of paper. Her purpose is to understand, on a very detailed level, exactly what it's made of.

The word "Martian" isn't a mistake, nor is it meant to be taken metaphorically. This rock, about the size of a potato and weighing a little over four pounds, comes from the planet Mars. It's one of twelve meteorites on Earth that planetary scientists are convinced came originally from the surface of the red planet. Their mineral and chemical composition is unlike anything on Earth or the Moon or even the asteroids and comets, where most meteorites came from. But they match nicely the chemistry of Mars, a chemistry that was studied directly during the Viking missions of the 1970s.

The meteorites come from Mars, but they were all found on Earth — in Egypt, France, India, and mostly Antarctica, which is where Thomas-Kaperta's specimen showed up. They dropped out of the sky, the victims of a highly improbable set of circumstances that was nevertheless repeated several times: Asteroid smashes into Mars; debris is blasted into interplanetary space; some eventually wanders into the gravitational influence of Earth and falls to the ground.

Naturally, these meteorites are valuable commodities. NASA is still years away from bringing back a single molecule from Mars for study, but here, in the form of these rocks from space, are pounds and pounds of Martian matter. Dozens of scientists in labs all over the world are looking at the Mars meteorites, performing every test they can possibly think of to understand what makes Mars tick. Thomas-Kaperta is one of them, and now she's ready to report on her findings.

What she's found inside ALH84001 is a bunch of tiny crystals—crystals so small that five million of them laid side by side would form a line just one inch long. The crystals are magnetite, a magnetic mineral made of oxygen and iron. Magnetite is reasonably common on Earth and quite familiar to geologists. It's also familiar to biologists. Magnetite forms naturally inside the bodies of certain living organisms, notably migratory birds and water-dwelling bacteria. Its function is navigational: Armed with magnetite, organisms can sense the direction and intensity of Earth's magnetic field and use the information to find their way to an old nesting ground or a new source of food. The birds and bacteria come equipped with an internal compass.

Magnetite can form outside living organisms as well, but when that happens, it doesn't tend to look like the tiny, cubical, and teardrop-shaped crystals that are typical of biological magnetite. The crystals in the ALH84001 are tiny cubes and teardrops.

That's not all Thomas-Kaperta has found. There's another magnetic mineral, a form of iron sulfide known as gregite, inside the meteorite. Like magnetite, gregite occurs inside the bodies of bacteria; like magnetite, it's there for use as a biological compass. It, too, can form nonbiologically, but again, the particles of gregite in Thomas-Kaperta's sample look precisely like what you'd expect to find inside a bacterium.

So in the abstract for her paper—the one-paragraph summary that helps colleagues figure out which presentations sound interesting—Thomas-Kaperta ended with: "This is suggestive of biogenic activity." She was saying it looked as though something had lived inside this meteorite—when it was still on Mars.

Now she's worried, and so are the other scientists working with her on this project. Thomas-Kaperta is about to go before her professional colleagues and suggest that she has evidence for life on Mars. What's worrying her is not the obvious, however. She isn't afraid she'll be laughed off the stage or lampooned in the media. Like Paul Butler and Geoff Marcy, Thomas-Kaperta is an extremely careful scientist. She second-guesses herself at every step, looks for ways she might be fooling herself. She knows that uttering the words "life on Mars" is something you don't ever want to do, even with the word "suggestive" in front of it, if you're not sure of what you have. But she also knows magnetite when she sees it, she knows gregite, and she knows that her conclusion is scientifically supportable.

Thomas-Kaperta is therefore not worried about her reputation; she's worried about her timing. Her microscopic searches are one part of a larger, on-going project to characterize this Mars meteorite, one line of detective work in an investigation that has proceeded on many fronts. The work has been going on for nearly two years, and it's finally starting to come together. If she steps forward now, she could blow the whole deal. The team is already nearly finished with a major research paper, pulling together the combined conclusions of nine scientists in what they believe will be a powerful case.

But it's crucial, they believe, to publish their paper in the most reputable scientific journal they can. Reasonable or not, their assertions will be widely challenged. They want every bit of credibility they can get on their side. So they're aiming for *Science*, the unassailably prestigious weekly journal of the American Association for the Advancement of Science. If any part of their paper appears somewhere else first, though—in the published proceedings of a Lunar and Planetary Science conference, for example, or in a newspaper or popular magazine—*Science* will declare it old news and refuse to publish it.

Thomas-Kaperta hasn't considered that problem before. Neither have her partners. Now that they do, they realize they have only one choice. Thomas-Kaperta contacts the conference organizers and withdraws her paper. Other scientists who notice the abstract in the program will note her withdrawal; some of them undoubtedly will wonder whether the research was shaky, or whether she yanked it because she had lost confidence in her results. It won't be long before they learn otherwise.

* ○ * ·

It's ten o'clock at night in Mountain View, California, when the phone rings in Seth Shostak's bedroom. He answers and hears the voice of his boss, Tom Pearson. Normally, the last person you want to find on the other end of the phone at that time of night is your boss. And, indeed, Pearson's news is disturbing, although it has nothing to do with Shostak's employment status. Seth Shostak works at the SETI Institute—SETI being an acronym for the Search for Extra-Terrestrial Intelligence. Pearson has just received a phone call from Australia where the institute's flagship program, known as Project Phoenix, is currently based.

Like all SETI programs, Phoenix is an attempt to listen for the radio signals broadcast by alien civilizations. No one knows whether such civilizations exist or whether they use radio waves for communication. But if they do and we aren't listening, we'll miss one of the most important discoveries in history. So we're listening—not spending a lot of money, since it's a long shot, but listening nevertheless.

If Paul Butler and Kathie Thomas-Kaperta need to be extremely careful about reaching premature conclusions and to be vigilant to the dangers of fooling themselves, Shostak and other SETI researchers need to be even more careful. It's one thing to claim prematurely that you've detected a planet or found evidence "suggestive" of submicroscopic life; it's quite another to say you've just heard from E.T. The former might earn you condescending sympathy from your colleagues, but the latter will make you a laughingstock. SETI is a perfectly legitimate branch of astronomy. It does, however, have an inherently high giggle factor that has to be kept constantly at a minimum.

That's why Project Phoenix was designed with multiple anti-giggle backup systems and why Shostak finds his boss's call so disturbing. The 210-foot radio telescope at Parkes Observatory in Australia, Phoenix's primary listening antenna, has picked up a promising signal. That's no big deal; all sorts of things can masquerade as an alien broadcast. A satellite passing overhead, a radar beam from a nearby airport, a radio broadcast from Sydney bouncing off the ionosphere at an unlucky angle, or any one of a hundred other human-generated signals will do.

When a signal does come in, the Phoenix computer tells a second radio telescope several hundred miles away to take a look. Most of the time it sees nothing; the signal is very local and clearly not alien. Once or twice a week, though, the second telescope sees it, too. That still doesn't prove anything; plenty of human signals have enough range to trigger two widely separated telescopes. When that happens, Phoenix scientists go to the next step in their self-debunking process: They aim the main radio dish away from whatever star they've been pointing at. If the signal is still coming in, it's not coming from the star and is thus almost certainly not coming from a distant solar system.

Tonight, though, the signal at Parkes has passed not only the first test— the backup telescope sees it, too—but also the second. When the telescope operators aim away from the star, the signal disappears. When they move it back, the signal comes back. Now they're trying a third test. In an hour the star will be setting below the horizon. If the signal winks out at the same instant, the coincidence will be too strong for any other conclusion: There is quite possibly a radio beacon aimed at Earth from another star system.

Shostak's wife is lying in bed reading, but he can't sit still and keeps pacing around the room. He has thought about the moment of contact for years—how he would feel if it really happened, what he would do, how he would deal with the knowledge that we are not alone in the universe. He has talked it over in late-night bull sessions with other SETI researchers. Now that the moment may finally be here, he's mostly feeling disoriented—and very, very nervous.

Finally, the hour is up. The phone rings. It's Tom Pearson again. The star has disappeared. The signal hasn't. They never will figure out what it was, but like all the other alarms SETI researchers have responded to over the past three decades, this one is false. But the next one may not be.

* O *

Step outside on a moonless, cloudless night and look up at the sky. If you're well away from any densely populated area, you can see several thousand stars without the help of a telescope or binoculars. Every one of them is a sun—and, that being the case, it's almost impossible not to wonder whether one of them has planets that are home to intelligent creatures looking up into their own night sky and wondering.

Trying to figure out our place in the universe, our standing in the cosmic scheme of things, is probably as old an activity as walking. The question is one which all religions attempt to answer, and much of science as well. The more specific question of whether other beings inhabit distant worlds is at least as old as recorded history. The ancient Greeks actively debated whether the universe is home to other worlds and other life, as did medieval clerics and Renaissance philosophers and nineteenth-century scientists.

By 1995 the question had not been settled. Common sense certainly favors the idea that life must be everywhere. The Milky Way contains billions of G-type, Sun-like stars. It's almost inconceivable that out of all these billions, ours is the only place where intelligence exists. But common sense has proven wrong in the past. If humanity is truly alone among the hundred billion stars of the Milky Way, which is itself one of tens of billions of galaxies in the universe, then the laws of Nature are indeed perverse—but nowhere is it written that they must be otherwise.

If, on the other hand, there is intelligent life on a significant number of other worlds, humanity is almost certainly not the most intelligent, the most accomplished, the most anything. The law of averages essentially guarantees that we are mediocre. That in turn leads to all sorts of speculations about what might happen if we ever come into contact with extraterrestrials. Perhaps we'd be in horrible danger. Perhaps we'd be able to tap into the eons-old knowledge of other galactic civilizations to learn the answers to questions we can't even formulate yet. Both possibilities have been thoroughly explored, of course, in science fiction books and movies. *E.T.* and *Independence Day* are perhaps the most memorable recent examples.

Without hard data, though, the question of humanity's place in the universe is purely a philosophical one. Unencumbered by the facts, you can take either side of the question and present convincing arguments both for and against the existence of intelligent alien life. And until 1995 there were essentially no facts. Nobody knew for sure that Sun-like stars had planets orbiting them, and without planets, it's hard to imagine life. Even if planets existed, nobody could say for sure whether the emergence of life, given the right conditions, was wildly improbable or nearly inevitable.

Yet after decades, centuries, millennia without actual information, it has suddenly and dramatically begun to flow. Paul Butler and Kathie Thomas-Kaperta and their collaborators and competitors are at last beginning to provide real data that other scientists can pore over. It isn't much so far. Butler is talking about a planet, not a galaxy filled with planets; Thomas-Kaperta is talking about microbes, not Vulcans like *Star Trek's* Mr. Spock. Inferring the existence of actual walking-around (oozing around?) aliens from these meager data requires a long chain of reasoning that could break at any point.

The chain of reasoning has a name. More than three decades ago, Frank Drake—then a young radio astronomer and now an eminent scientist and also the president of the SETI Institute—boiled the whole thing down to an equation that became known as the Drake equation. In prose form the equation says that the number of civilizations theoretically detectable at a given time depends on how many stars there are, how many of those have planets, how many of those are hospitable to life, how many of those give rise to life, how many of those give rise to technological civilizations, and, finally, how many of those survive their own success. This last point is extremely important. If the average high-tech civilization blows itself up after fifty or one hundred years, we probably won't ever get to talk to anyone.

The Drake equation is so concise and so apt a statement of the extraterrestrial intelligence problem that nobody has ever come up with a better way of stating it. Researchers in a wide range of fields—biologists thinking about the origin of life, astronomers trying to tease the faint signature of other worlds out of the motions of nearby stars, theoretical astrophysicists trying to unravel the mysteries of planet formation—can and often do describe their work as an attempt to crack part of the Drake equation.

In that sense, Butler's and Thomas-Kaperta's research, though preliminary, is extraordinarily important. Butler's observations don't yet tell astronomers how many stars have planets around them, but they at least prove that some beyond the Earth do. Thomas-Kaperta's work doesn't prove that life is common throughout the Milky Way, but it makes that proposition far more convincing.

Not everyone will be convinced, of course, even by the most rigorous scientific research. Some astronomers will suggest that Butler's planets aren't really planets after all; many biologists will question whether Thomas-Kaperta's team has found evidence of life or simply of something that looks like life. Dealing with their critics will probably take as much time as the original research, and more. "That's what we expected," says Everett Gibson, a NASA geologist who works with Thomas-Kaperta. "That's the way science is supposed to work. You go out there with your best evidence and hope that your critics haven't thought of something you didn't anticipate."

While the critics take their best shots at undermining these first apparent breakthroughs in solving the Drake equation, the scientists working on planet searches and microbes from Mars and SETI programs and a dozen other related research programs are moving on. The discoveries of Butler and Thomas-Kaperta and their colleagues are a beginning, not an end. Even as the scientists announced their shocking results to an astonished world, they were digging deeper into their data and trying to gather more. Before the year is over, Butler and Marcy will have more than 70 Virginis to talk about, and they'll take their observing program to the most powerful telescopes on Earth to probe deeper into the Drake equation. By early in 1997, Thomas-Kaperta's group will claim new, more powerful evidence for life in their Martian meteorite.

And they won't be alone. Marcy and Butler are the most prolific planet-hunters in the world, but they are not the only ones. Thomas-Kaperta's group has learned more than anyone else about what's hidden inside the Martian meteorite ALH84001, but ALH84001 is not the only Martian meteorite on Earth, and hers is not the only team looking for signs of life. In science as in sports, proving that a difficult feat is nevertheless possible makes it easier, suddenly, for others to repeat it.

Within a few months after Butler's startling discovery, that is precisely what will happen, and Drake-equation science will become the focus of just about every major astronomical and general science conference. While the astronomers talk of new planets and the meteorite analysts discuss the fine structure of Martian rocks, biologists will offer their own insights into the origin of life; planetary scientists will find powerful new justification for sending space probes back to Mars and even out to the moons of Jupiter; and instrument designers will be laying out their plans for giant new telescopes that will make the Hubble Space Telescope and the Keck Observatory look small and old-fashioned. Today, the best Paul Butler can do is see evidence that a star is wobbling; within a decade or two he may be able to see Earth-like planets directly and even probe their atmospheres for telltale signs of life.

SETI researchers like Seth Shostak, meanwhile, will be looking to the skies with new encouragement: There *are* planets out there after all, and there may be, or may once have been, life on Mars. So they'll be continuing their own campaign to make the planet-hunters' work obsolete. They will bring Project Phoenix from Australia to a new, semipermanent home at the National Radio Astronomy Observatory in Green Bank, West Virginia, and help fund a handful of smaller projects around the world.

And while they assume their search will take years at least, and probably decades, it's also true that an alien radio broadcast could be sweeping across Earth as these words are being written or being read. Long before a post–Hubble Space Telescope takes pictures of an alien Earth, the question of whether other civilizations populate the universe may be solved in one single, dramatic message from the stars. Either way, the ancient mystery of life in the universe has finally, after uncounted thousands of years, begun a transformation from a religious, philosophical, purely intellectual question to a scientific one. Soon—quite possibly within a decade or two—it may be solved.

Chapter 1

HISTORY

The corridor is dark, the air damp and bitterly cold. A powerful odor of mildew mingles with the smell of human waste. You can hear water dripping somewhere behind the walls. Rats scurry across the floor, searching for crumbs. These are the dungeons of the Roman Inquisition. The date is February 8, 1600. Inside one of the cells, a man sits in silent contemplation. His name is Giordano Bruno, and he has just been sentenced to burn at the stake ten days hence, on the orders of Pope Clement VIII.

Bruno has been condemned as a heretic, but his crime is not so much that he espouses forbidden beliefs as that he can't keep his mouth shut about them. He is basically a pain in the neck. He has all these crazy ideas, and he simply won't stop talking about them, even when they're obviously irritating to the Church. It doesn't seem to matter which church, either: Bruno is an equal-opportunity pain. In the thirty-five years that have elapsed since he began his career as a Catholic seminarian in Naples, Bruno has been excommunicated by the Catholics, the Calvinists, and the Lutherans. He has wandered from Italy to Geneva to Paris to London to Germany spreading his heresies, and finally back to Italy. Not to Rome, of course—he may be outspoken, but he's not suicidal. Venice, though, is a relatively liberal town. They don't generally sentence you to death here for espousing unpopular ideas.

But Bruno managed to antagonize his Venetian patron, the aristocrat Giovanni Mocenigo, who originally invited Bruno here to teach him memory tricks—for besides being a cleric and a philosopher and a magician, Bruno is a renowned memory expert. Maybe the lessons didn't go so well, or perhaps Mocenigo was infuriated at learning of Bruno's plans to skip town and head back to Germany. Whatever the reason, Mocenigo denounced him to the Venetian Inquisition, and once Bruno was in the Church's hands, Rome would arrange his extradition.

The trial in Rome lasted seven years—long periods of languishing in the dungeons punctuated by sessions of intense, oppressive interrogation. He recanted his beliefs, then recanted his recantation. Finally, the Pope said "enough" and demanded Bruno's execution. "Perhaps," said a defiant Bruno after his sentence was formally read in court, "your fear in passing judgment on me is greater than mine in receiving it."

Ten days later, Bruno is bound, gagged, and hauled off to the Campo di Fiori, one of the city's public squares. He is tied to a stake and, with all due ceremony, burned to a crisp. Witnesses will later claim that as the flames blazed up around him, someone tried to press a crucifix into his hands, but he refused to take it.

Bruno could probably have been sentenced to die for any number of his beliefs. He thought, for example, that the true religion was not orthodox Christianity but, rather, a mystical quasi-Christianity handed down from the ancient Egyptians—not the sort of thing the Pope would have appreciated. But the Church was officially offended by the cosmology Bruno embraced. He believed, in direct opposition to the official Catholic position, that the Earth was not alone in the universe. In his 1584 work *De l'infinito universo e mondi* (On the Infinite Universe and Worlds), he wrote:

There are countless suns and countless earths all rotating around their suns in exactly the same way as the
seven planets of our system. We see only the suns because they are the largest bodies and are luminous, but
their planets remain invisible to us because they are
smaller and non-luminous. The countless worlds in the
universe are no worse and no less inhabited than our
Earth. . . . Destroy the theories that the Earth is the center of the Universe!

Bruno is seen today as a visionary, the secular patron saint of scientists
who are searching for life on other worlds. In fact, his ideas came from philosophy, not science. The universe must be filled with other worlds, he insisted, because reason dictates that it must be so. He had no actual evidence
for the existence of such worlds. Neither, on the opposite side of the argument, did the Church. Its opposition to Bruno and others like him was
purely philosophical as well. Multiple worlds didn't fit into Catholic theol-
ogy; therefore, such thinking was wrong.

Neither of these worldviews was new in 1600. In Western civilization, the
idea that the universe is home to other worlds dates back at least to the
Greek philosopher Thales, who lived in the seventh century B.C., and to his
student Anaximander. (Thales, in turn, may have gotten the idea from the
Egyptians.) In the fifth and fourth centuries B.C., the philosophers Democri-
tus, Epicurus, and Leucippus took up the many-worlds banner, arguing that
the universe was infinite in extent and that it must encompass an infinity of
populated worlds. These three philosophers were known as the atomists because they believed, nearly twenty-five centuries before modern physics
reached the same conclusion, that the cosmos is made up of infinitesimal,
indivisible particles they called "atoms."

The atomists hadn't a prayer of proving that either atoms or infinite
worlds existed. The former were too small, and the latter were much too far
away—well beyond the stars, certainly, for the "other worlds" the atomists
believed in would more accurately be described as other universes. But test-

ing a theory with evidence in the modern sense was not standard procedure in ancient Greece. Atomism was a philosophical viewpoint, not a scientific theory, and it would stand or fall on its intellectual persuasiveness rather than on experimental confirmation.

As it turned out, atomism wasn't persuasive enough. It lost out in the fourth century B.C. to Aristotle's more intuitively obvious rival cosmology. There is one world, Aristotle argued—the Earth. It lies at the center of the universe, and around it revolve invisible heavenly spheres, nested one inside the other like the layers of an onion. Embedded in individual spheres are the Sun, the Moon, Mercury, Venus, Mars, Jupiter, and Saturn (none of which was considered a "world"; they were just planets, from a Greek word meaning "wanderer"). In the outermost sphere rode the stars. It made perfect sense.

This comfortable and utterly incorrect view of the universe was passed on from the Greeks to their intellectual descendants, the Romans. When Rome became a Christian empire, the Earth-centered cosmology was retained, since it fit nicely into the cosmos described in Genesis.

So matters remained right up through the Middle Ages. There was just one world, according to Christian dogma, which rendered talk about other worlds pointless except as a purely philosophical speculation. A few scholars, particularly a group known as the Scholastics, dabbled with the idea that God could in principle have created other worlds without strictly violating Aristotle's system. But as the astronomer and historian Steven J. Dick writes in his 1996 book *The Biological Universe* (a thorough and highly readable history of the question of life in the cosmos), "While [other worlds] were possible, all the medieval Scholastics stressed, God in reality had not created more than one world." That is to say, God could have done it if he had wanted to. He just didn't want to.

Besides, this talk of other worlds was purely academic. Where were they? A world was a vast area of land and sea and mountains sitting at the center of the cosmos. It wasn't some pinprick of light out in the sky.

For all its elegance, however, Aristotle's model of the universe didn't match the observed facts. It was well known that the planets did *not* move at uniform speed across the sky, as they should if they were riding on rotating crystal spheres. Mars, Jupiter, and Saturn, for example, were seen routinely to reverse direction, then start moving forward again. (Now we know it's just an optical illusion: The Earth, moving in a tighter, quicker orbit, catches up with and passes these planets, making them seem to move backward.)

The philosopher Ptolemy dealt with that problem in the second century by proposing that the celestial spheres had smaller spheres attached to them and that the planets were embedded in these smaller spheres. The motions of spheres upon spheres allowed for backward motion and thus came closer to producing the actual motions of the planets. The system was somewhat unwieldy and inelegant, though. These smaller spheres were known as "epicycles," and scientists still use the term when ridiculing awkward additions created in desperation to make an incorrect theory work.

In 1543 a better idea was put forward by Nicolaus Copernicus, a Polish astronomer and cleric. Copernicus realized that many of the complications of the Aristotelian/Ptolemaic cosmos could be eliminated simply by assuming that the Earth and planets orbited the Sun. He wasn't the first to suggest this arrangement. As with so many scientific ideas, the Greeks had been there first. The philosopher Heraclitus had proposed the same notion, but like that of atoms, it didn't prevail.

Revolutionary as it was, Copernicus's cosmology required epicycles, too, because he believed heavenly orbits were perfectly circular. They are, in fact, elliptical—slightly oval, as Johannes Kepler later realized. Even so, the model was far simpler and more elegant than the mess astronomers had been working with for more than a millennium.

Elegant or not, demoting the Earth from the center of the universe, in defiance of the Church's teaching, could have gotten Copernicus in big trouble, as it would Bruno a few decades later. Fortunately (so to speak), Copernicus was near death from a massive stroke when his radical new theory was published in the book *De revolutionibus orbium caelestium* (On the Revolutions of the Celestial Spheres). The Church couldn't do much to him.

Neither, however, could Copernicus do much about Andreas Osiander, a Lutheran who was supervising the printing process for the dying philosopher. Osiander, horrified at what Copernicus's theory implied, took it upon himself to add a preface to the work. It stated that the Sun-centered universe was really a mathematical fiction, not a real description of the cosmos, and was only meant for use in simplifying calculations. Osiander somehow forgot to sign his name to the preface, suggesting falsely that this was Copernicus's own view. Copernicus died the same day that proofs for the book were printed; it is not known whether he ever even knew of the preface's existence.

If Copernicus had lived on and had survived the attention of the Church, he still might have dropped dead of heart failure when Giordano Bruno came along with his crazy talk of other worlds filled with intelligent beings, stretching out to infinity. Copernicus would certainly have been fascinated, though, with the new picture of the heavens that Galileo would draw less than a decade after Bruno's execution. Contrary to popular myth, Galileo didn't invent the telescope, which was the creation of Dutch opticians. But he was among the first to turn the new device on the night sky, guaranteeing himself a place in history for having the presence of mind to put on paper what he saw.

What Galileo saw made it much harder to dismiss the Copernican cosmos, as scholars other than Bruno had generally done. He saw, for example, that the Moon is not a perfect sphere, as Aristotle had claimed, but instead is pocked with craters and wrinkled with mountain ranges. He saw that Jupiter has its own set of moons, contradicting Aristotle's insistence that everything in the heavens revolved directly around Earth. He saw two pronounced bulges on Saturn, which he presumed were also moons (his telescope wasn't sharp enough to show the planet's huge rings as distinct objects). And he saw that Venus goes through phases like the Moon, in a pattern that the Aristotelian/Ptolemaic system could not possibly explain.

The Church was not pleased. It became actively displeased when Galileo published a treatise entitled A Dialogue on the Two World Systems, which essentially suggested that anyone who believed in an Earth-centered cosmology was a fool. But at least Galileo was no Bruno; there was no nonsense about life on other worlds. Besides, Galileo recanted his beliefs during his own visit to the Inquisition. His sentence of life imprisonment was reduced to permanent house arrest.

By now, though, Copernicans were popping up everywhere. Armed with telescopes, astronomers were able to measure planetary motions with greater precision than ever, and they were finding that the Earth-centered, epicycle-laden cosmology was simply insupportable. Finally, in 1609, after a decade of work, the German astronomer Johannes Kepler realized that Mars's orbit was just slightly elliptical, not circular, a result he later generalized to the other planets. His model matched the behavior of the actual solar system. The Vatican kept Copernicus's works on its forbidden list until well into the 1800s (and didn't officially pardon Galileo until 1992), but the evidence that Earth was just another planet orbiting the Sun quickly became impossible to deny. By the late 1600s, the Earth-centered universe was essentially finished.

If the Earth was just a planet, though, then why couldn't the planets be Earths, with animals and even people living on them? Kepler was convinced they could be. He wrote a tract entitled *Somnium seu astronomia lunari* (Dream, or Lunar Astronomy) in which he suggested that the Moon was probably inhabited. He also wrote that since Jupiter's moons had been invisible to humans until lately, their only purpose could be to brighten up the night skies for indigenous Jovians. Galileo was more cautious; he would only go as far as suggesting that the dark blotches on the Moon were oceans. But others, including an English bishop, John Wilkins, and Pierre Borel, a French courtier, suggested not only that beings lived on the Moon but that earthlings would one day travel there to meet them. And Bernard Le Bovier de Fontenelle, in his 1686 book *Entretiens sur la pluralité des mondes* (Conversations on the Plurality of Worlds), argued boldly that every planet in the solar system is inhabited.

De Fontenelle's lively, readable book, which was wildly successful across Europe, also advanced the theories of his mentor, René Descartes, who had suggested some forty years earlier that every star is a sun and that every sun has its own retinue of planets. Although this sounds pretty much like Bruno's claims, it was based not on mysticism and philosophical deduction but on Descartes's idea, derived from observations of winds and water, that matter in the universe must gather itself into whirling vortices. This was a remarkably precocious proposal, as it would turn out. If the universe was filled with planets, then you didn't have to depend on only the other five then known to exist in our solar system; Jupiter, Saturn, Venus, Mars, and Mercury could be barren, and you'd still have lots of places where life might exist.

With these two developments—the suggestion that life might exist on more than one planet and the argument, based on physical theory, that planets are orbiting stars beyond the Sun—the modern approach to the search for life on other worlds, which would be formalized a little more than three hundred years later in the Drake equation, was already starting to take shape. If one couldn't yet find otherworldly life directly, one could at least estimate its likelihood by breaking the problem down into more manageable parts. One question might be "Is it reasonable to think there's life on the Moon or Mars or Jupiter?" Another would be "Are there Earths, Jupiters, Marses, and the like, circling other Suns?"

On the latter question, the conventional wisdom swung back and forth for centuries. There was no way to answer it directly. As Bruno had presciently written, "We see only the suns because they are the largest bodies and are luminous, but their planets remain invisible to us because they are smaller and non-luminous." The best astronomers could do was entertain theories about how planets might form, and line up behind the most persuasive. If you believed in a theory that formed planets through a series of low-probability coincidences, you wouldn't hold out much hope for the existence of other worlds. If you signed on to a theory that made solar systems at the drop of a hat, you'd have to believe that planets are everywhere.

Descartes suggested, for example, that matter was sucked into whirlpools that evolved into planetary systems. Since these whirlpools could occur everywhere, so could planets. In 1745 the Frenchman George-Louis Leclerc proposed instead that a comet had whacked into the Sun and dislodged chunks that became the planets. Collisions like this would presumably be rare, chance events, so if Leclerc was right, planets should be exceptional.

For the next three hundred years, scientific opinion oscillated between these two ideas, each of which was periodically refined and updated. In the 1750s, for example, Immanuel Kant improved Descartes's proposal, arguing that blobs of primordial matter had condensed under the newly identified force of gravity to form solar systems. And in the late 1700s, Pierre-Simon de Laplace, a French mathematician, refined Kant in turn, arguing that these blobs would necessarily flatten into disk-shaped clouds as they collapsed and then went spinning. At the center of such a cloud, where the blob was densest, a sun would form; farther out, the disk would fragment into rings, which would then condense into planets. Leclerc's theory was updated as well: Comets are obviously too small to affect the Sun, so the rogue comet was replaced with a rogue star smashing into the Sun, and direct impact was downgraded to a close encounter.

By the middle of the nineteenth century, astronomers were leaning heavily toward the Descartes/Kant/Laplace theory of planetary formation. This was partly because they wanted to believe that there were lots of planets in the universe, and partly because they thought they'd caught planetary systems in the very act of formation. The sky was studded with tiny hazy blobs of light, known as spiral nebulae, that looked suspiciously like Laplace's condensing disks.

By the beginning of the twentieth century, the pendulum had swung the other way. For one thing, astronomers realized that the spiral nebulae were actually galaxies like the Milky Way—huge collections of stars very far away rather than young solar systems lying nearby. For another, observers had discovered to their astonishment that more than half of the stars they looked at through their increasingly powerful telescopes were actually double-star systems, and some were triples or even quadruples. A solar system would have a hard time forming, they thought, if the young planets had to dodge all those stars.

Another problem was that a solar system should presumably retain the spin energy of the blob it originally condensed from. Physicists call this energy angular momentum; it's a function of an object's mass, size, and rotational speed. The laws of physics say that the total amount of angular momentum in any system has to remain constant, so if you change one factor, another must adjust to compensate. If the system shrinks in size—as a primordial cloud of gas would have to do in order to form a solar system—it has to spin faster to compensate for the shrinkage. Always cited to illustrate this phenomenon is the analogy of a spinning figure skater: The skater goes into a spin with arms extended and then, by doing nothing more than bringing his arms close to his body, spins faster.

According to this rule, the angular momentum of the solar system didn't make sense. The planets were orbiting too fast—or, conversely, the Sun was rotating too slowly—for the two to have condensed from the same cloud.

With all this evidence piling up against it, the condensing blob theory was looking more implausible all the time. By the 1920s, astronomers led by the British theorist James Jeans were convinced that close encounters between stars were, after all, the force behind the creation of planets. Based on the huge distances between stars, they judged, such close encounters should happen once every thirty billion years or so. One solar system could conceivably exist against the odds, but two? No way.

But the pendulum of scientific opinion swung again. Jeans's close encounter theory, while more detailed and convincing than any before, still had fatal flaws. He couldn't really explain the solar system's angular momentum, either. Beyond that, while a passing star could skim hot gas from the Sun, that gas wouldn't condense into planets; it would simply disperse. Lyman Spitzer, the scientist who proved this point in 1939, was a graduate student working under the legendary Henry Norris Russell when he wrote the paper that helped torpedo Jeans's theory.

While Spitzer, Russell, and other theorists were raising objections to Jeans's close encounter model, others were busy showing that reports of the condensing blob theory's death had been greatly exaggerated. The spiral nebulae were indeed faraway galaxies, but observers had also detected enormous clouds of gas floating within the Milky Way, which could plausibly fragment into blobs and collapse someday. The majority of stars were indeed part of multiple-star systems, but many were not—and, in fact, the same gases that might have condensed into a second star were the perfect construction material for planets when that didn't happen.

Even the angular momentum problem wasn't such a problem anymore: Astrophysicists now realized that the Sun would have spewed out a powerful stream of particles in the first million or so years of its life, a hurricane that would later weaken into the relatively gentle solar wind that exists today. As the analogy had it, this escaping gas would serve as the arms of a spinning figure skater opening wide. The Sun, shedding its own angular momentum, would slow down; the still-forming planets would not.

Angular momentum wasn't just vanishing as a potential difficulty with planet formation, it was being rallied in support of the process. In the mid-

1940s, University of Chicago astronomer Otto Struve noted that massive hot stars tend to rotate rapidly, while smaller, more Sun-like stars rotate quite slowly. It wasn't that the hot stars had condensed from bigger gas clouds, which had more angular momentum to begin with, because even when he allowed for that difference, the numbers didn't add up. Struve suggested that the difference could be explained by planets: The big stars had sucked in *all* the matter and all the momentum from their parent clouds (they were like the skater who pulls his arms inward), while the smaller stars had only been able to snare *most* of it, the balance being retained in the orbits of the planets.

By the end of the 1940s, the weight of theoretical argument and observational evidence had convinced just about everyone that the impact theory was dead and that solar systems were probably sprinkled throughout the Milky Way. There was no guarantee that the planets around other stars harbored life, but probably there were places where life could perch if it did arise.

Within our own solar system there are, of course, lots of such places, as Kepler, Wilkins, Borel, and de Fontenelle had all noted in the 1600s. As the theory of planet formation evolved over the centuries, so, in parallel, did the local search for extraterrestrial life.

It took no time at all for astronomers to rule out most of the solar system as inhospitable. The Moon, they discovered, has no atmosphere to speak of. This was clear when they watched through a telescope as it passed in front of a background star. If there were air on the Moon, the star would first dim as it was blurred by the atmosphere, and then it would disappear behind the edge of the Moon itself. What actually happens is that the star just winks out, without any preliminaries. As for the planets, Mercury was much too close to the Sun and thus too hot, Jupiter was too far away and too cold, and Saturn was farther still. Uranus, Neptune, and Pluto, when they were discovered in 1781, 1846, and 1930, respectively, were even more distant and colder. Astronomers were convinced by the turn of the century that the only places where life might conceivably exist were Mars and Venus.

About Venus, scientists could only speculate. The planet was perpetually fogged in, covered by a permanent layer of light-reflecting clouds. These clouds make Venus the brightest object in the sky after the Sun and the Moon, but they prevent anyone from getting even a glimpse at its surface. Perhaps Venus was covered with a planet-wide ocean; perhaps it was dominated by swampy rain forests, with dinosaurs roaming a world stuck in an unending Jurassic period. There was no way to tell. By the 1930s, though, astronomers did know that there wasn't any oxygen to speak of in Venus's atmosphere and that the surface was at the boiling point of water, at least. Interest in Venus as a living planet ceased after that.

But Mars was a planet with some potential. In the late 1700s, telescopes revealed that the red planet had polar ice caps, just as the Earth does, and that they changed in size with the seasons, as ours do. Other markings on the planet—dark splotches that swam in and out of view as Mars shimmered in the blur of Earth's atmosphere—seemed to change their size and shape as well. These could presumably be vegetation, growing during the Martian summer and dying in winter.

Finally, in 1877, when the orbits of Earth and Mars brought the two planets closer together, an Italian astronomer named Giovanni Schiaparelli made an astonishing discovery: The Martian surface was crisscrossed with a network of straight lines. Not every astronomer who looked could see the lines, but many confirmed the discovery. Schiaparelli saw them when he looked again in 1879 and again in 1882. He called these lines *canali*, or channels, a term that suggests they might be natural features—and, indeed, Schiaparelli thought they probably were, though they seemed so perfectly straight that he didn't dismiss the possibility they were constructed by Martians instead. "I am very careful not to combat this supposition," he wrote, "which includes nothing impossible."

When Schiaparelli's writings were translated into English, though, the word *canali* became "canals," which suggested strongly that they were artificial. That suited the prejudices of many astronomers just fine, and piqued the interest of a wealthy Bostonian and Harvard graduate named Percival Lowell. Lowell was captivated by Schiaparelli's work. He'd evidently learned in 1893, moreover, that the Italian's eyesight was starting to fail, and he was determined to carry on Schiaparelli's work himself.

Lowell hooked up with William Pickering, a Harvard astronomer, and agreed to finance a scientific expedition to Arizona where the skies were much clearer than they were in damp Boston. The expedition quickly escalated into a permanent observatory, paid for and used by Lowell to supply evidence for what he already believed to be true—that there were intelligent beings on Mars.

By 1895, Lowell had already published a popular book on the topic, setting forth his arguments. First, he said, there were no oceans visible on the planet, so it must be dry. If there were intelligent Martians, they'd probably have to dig canals to bring water from the poles down to their farms. Second, there were lines—just what canals should look like from space. (Lowell actually thought he was seeing vegetation growing alongside the canals, marking their path just as cottonwood trees mark the course of a stream in the desert.) Third, the canals led to the dark blotches that waxed and waned over time, so these must be farmlands.

This reasoning was rejected by many scientists partly on the basis that having a particular answer in mind before you start an experiment makes it just about impossible to be objective. They were also rather indignant that Lowell had published a popular book, going direct to the public rather than to the astronomical community within a year after starting his observations. Schiaparelli, by contrast, observed Mars for years and wrote mostly technical papers aimed at his colleagues. And unlike Lowell, he had been very careful not to jump to any conclusions about what the canals actually were.

But other astronomers believed the canals might be real, and the general public, uninhibited by scientific caution, was easily convinced. It wasn't just the canals that contributed to the widespread belief that Mars was home to intelligent beings: In 1897, H. G. Wells wrote *The War of the Worlds*, which featured an attack by malevolent Martians; in 1917, Edgar Rice Burroughs wrote the first installment of the long-running John Carter of Mars series.

In 1899 the inventor and engineer Nikola Tesla noted some electric noise in his power transmission apparatus that he attributed to interplanetary signals. And in 1920, Guglielmo Marconi reported that his receivers had been picking up mysterious radio signals that he could track to no known transmitter. He thought the signals could conceivably have come from Mars; over the next few years Marconi would detect many more such signals and become convinced that the Martians were indeed calling.

In response to these reports (which were disputed by many scientists, including Albert Einstein), proposals about how best to open a conversation with Mars began to emerge. The Sperry Gyroscope Company suggested it could aim two hundred or so of its high-intensity searchlights, originally developed to spot enemy planes in World War I, at the red planet. Others suggested setting immense fires in the middle of the Sahara desert to catch the Martians' attention. The French Academy of Sciences offered a reward of one hundred thousand francs to the first individual who sent and received a message from another planet.

Although there is no evidence that he was trying for the French prize, the American astronomer David Todd was especially creative about talking to Mars. He suggested sending up powerful transmitters in a hot-air balloon, to try to communicate with less of Earth's atmosphere in the way. He came up with a scheme for building a huge telescope with a mine shaft for a tube and a mirror of spinning liquid mercury fifty feet across at the bottom to take close-up photographs of the planet when it was directly overhead. And in 1924, when Mars would be making a particularly close approach to Earth, Todd proposed that all of the world's radio transmitters go silent so that any Martian signals could get through without interference.

The last scheme was the only one actually carried out—sort of; only a very few transmitters did shut down at the appointed time. But the U.S. Army and Navy both issued orders to their radio operators to avoid unnecessary transmissions and to listen for anything unusual. The Army Signal Corps even ordered a code specialist to stand by to decipher the message if it came in. It didn't—at least not in any form that couldn't be attributed to natural causes.

In the end, the furor died away. Nobody ever did hear the Martians transmitting. By the 1930s, the canals were gone as well. Astronomers now had much more powerful instruments than the eight-inch telescope Schiaparelli had used, or Lowell's twenty-four-inch. (The size refers to the diameter of the lens or mirror that gathers and concentrates light; a bigger diameter means more light-gathering power.) A Greek astronomer, Eugene Antoniadi, had looked through a thirty-three-inch telescope and seen various blotches, with edges of contrast between light and dark areas that the human eye might interpret as thin straight lines if looking through a modest telescope through the blurry atmosphere. The lines were an optical illusion. American astronomers had used the more powerful sixty-inch telescope at Mount Wilson in southern California and, after it was built in 1917, Mount Wilson's one-hundred-inch telescope, and had come to the same conclusion.

Even these disappointments could not destroy the idea that life of some sort, even intelligent life, might exist on Mars. Orson Welles's 1938 *War of the Worlds* radio play, which sent thousands of Americans fleeing their homes in panic and kept many more glued to their radios in terror, was clear evidence that Martians still lurked disturbingly in the back of the public mind. Observations of the planet's atmosphere in the 1940s showed that Mars had very little oxygen, which finally ruled out animal life of any sort and presumably intelligent life as well. But the dark patches were still there, and they still changed shape over the years. Perhaps this simply marked the ebb and flow of giant dust storms that hid different parts of the surface, as a young Carl Sagan suggested in the late 1950s. Still, as late as the early 1960s, holdout scientists were arguing that the patches might be primitive plant life—lichenlike organisms, perhaps—coating the surface rocks.

The majority, though, were convinced that if the red planet harbored any life at all, it must be bacterial at best. But finding even the tiniest germ on Mars would be a terrifically exciting discovery. It would be a powerful indication that Bruno was probably right, for if life could arise on two different planets in a single solar system, it must be ubiquitous—and some of it must have progressed to the point of intelligence. After three and a half centuries of wrestling with their theories, most astronomers now believed that the Milky Way was littered with planets. The universe was packed with real estate where intelligent creatures might be flourishing.

But while these presumable planets would give aliens a place to live, there was no assurance that the aliens actually existed. Astronomers had plenty of opinions on this question, but their opinions weren't worth much, and they knew it. They also knew that biologists, biochemists, and organic chemists, whose sciences had matured much later than astronomy, finally did have useful things to say about life on other worlds.

To begin with, Charles Darwin's theory of evolution by natural selection, proposed in 1849 and refined over the next hundred years, provided a plausible mechanism for the development of life not only on Earth but, by implication, anywhere in the universe. You could start with an utterly primitive single-celled organism and, given enough time, evolve dinosaurs and dolphins and chickens and human beings—or whatever the alien equivalents might be. If the laws of nature managed to get as far as creating bacteria on a hospitable planet, the rest was easy.

What was unclear still was how easy it was to make life in the first place. Scientists had been unraveling the basic biochemistry of life since the 1920s. They hadn't been focused on the question of extraterrestrial life. But the astronomers knew that if biologists and chemists could figure out what made life work, they might also be able to explain how it first arose—and thus offer some insight into how likely it was for there to be life on other worlds.

By the late 1950s, biology had plenty to say on all these questions. It was clear that life was based on carbon, oxygen, hydrogen, and nitrogen, combining to form amino acids, which linked together to form proteins and enzymes and lipids and polysaccharides and nucleic acids—the building blocks and information-carrying molecules of living cells. It was plausible, at least, that simple chemicals like methane, carbon dioxide, and ammonia— chemicals that may well have existed on primitive Earth—could have been forced, by incoming ultraviolet light from the Sun or electricity from lightning on Earth, to form more complex chemicals.

In the early 1950s, Melvin Calvin at the University of California and Stanley Miller of the University of Chicago did lab experiments that proved the point: Calvin used a cyclotron to irradiate carbon dioxide and water vapor, and produced formaldehyde and formic acid; Miller used electricity to transform a mix of gases into amino acids. This was a long way from creating life and, thus, a long way from saying anything definitive about how easy or difficult it might be for life to arise on the average planet. Still, Miller's and Calvin's experiments at least proved that life elsewhere was not implausible.

By the end of the 1950s, the same was true for the broader question of life in the universe. Astronomers hadn't actually found planets around other stars, but the prevailing theory said the planets should be there. Intelligent life apparently didn't exist on Mars, but bacterial life might, and with that hope came the implication of a galaxy filled with biology.

So while the suicidal propositions of Giordano Bruno were in one sense still as speculative as they had been in 1600, they were now based on science and not simply on philosophical deduction. Threads of seemingly unrelated research were being woven into a sort of tapestry. The threads would continue to remain distinct: Astronomers would refine their theories of how solar systems formed and would finally have instruments that let them actually find extrasolar planets. Planetary scientists would begin to understand the past history and current conditions, including how hospitable it would be to life, on the planets and moons of our own solar system. Biologists would better understand how and why life began here on Earth.

And unlike at any time in the past, these scientists would begin talking to each other. Their orbits would intersect. They would keep each other informed of the latest thinking in their own fields. While they remained astronomers, biologists, and chemists, they would also start to think of themselves as practitioners of a new, multidisciplinary field known variously as exobiology, astrobiology, or bioastronomy, a coherent if loosely organized research program in search of alien life. Starting at about the time of manned space exploration, the search for life began its final transition from philosophy and theory into hard, experimental science.

Chapter 2

THE DRAKE EQUATION

Einstein's equation $E = Mc^2$ is the most famous equation of the twentieth century, if not of all history. The second most important equation of the century is more cumbersome but easier to understand. Frank Drake's equation reads $N = R^* F_p N_e F_l F_i F_c L$. It has taken on almost talismanic quality because it addresses a question that most people care about a lot more than they do the curvature of space-time. In English, the equation states that N (the number of detectable civilizations) equals R^* (the rate at which Sun-like stars form) times F_p (the fraction of stars that form planets) times N_e (the number of planets per solar system hospitable to life) times F_l (the fraction of planets where life emerges) times F_i (the fraction of life-bearing planets where intelligence evolves) times F_c (the fraction of such planets where the inhabitants develop interstellar communication) times L (the length of time such civilizations continue to communicate before they lose interest or blow themselves to atoms or succumb to natural causes).

We already know that N = 1 at a minimum, because ours is such a civilization. We have the technology to broadcast a signal to nearby stars that another civilization, no more advanced than our own, could detect.

But it is the belief and hope in N = 2, or a higher number, that has fired the public imagination. *Star Trek, Independence Day, Star Wars, E.T., Close Encounters of the Third Kind,* and *Contact* have all been smash hits because most of us want to believe that humanity is not alone in the universe. We're convinced aliens exist. The Drake equation tells us that scientists hope so, too, and that they consider the subject legitimate enough to formalize into the language of mathematics. Equations are things that professors scrawl on blackboards or put into textbooks. Equations are a signal that this is real, not science fiction.

Drake's equation sends an equally powerful signal to scientists. It tells them that the vague, still highly speculative subject of searching for extraterrestrial intelligence can be broken down into discrete, manageable steps. They had been proceeding on this assumption already, of course—the planetary scientists thinking about places for E.T.s to live, the biologists thinking about the origin and evolution of life. Others had even said explicitly that these different elements needed to be considered together in order to think about the probability of life on other worlds. But the proposition had never been so clearly stated or so neatly tied together before the day in October 1961 when a tall, skinny, prematurely white-haired young radio astronomer stood up at a scientific meeting in West Virginia and scrawled his equation on a blackboard.

Frank Drake had no idea at the time that he was making history. He'd come up with his equation only a few days earlier, and not because he felt compelled to formalize an entire area of scientific investigation. Drake invented the equation because eleven people were about to arrive for a conference he had organized on the search for life on other worlds, and he needed something for them to talk about.

Extraterrestrials had been on Frank Drake's mind for many years at that point. He'd first begun thinking about the possibility of life on other worlds as a teenager, largely in reaction to the strict Baptist Sunday school he attended while growing up in Chicago. The Hyde Park Baptist Church was on the grounds of the University of Chicago, and Drake's teachers included faculty members from the university's Oriental Institute. They would take Drake and his classmates over to the institute to look at mummies and Egyptian artifacts, to impress on them that the stories in the Judeo-Christian Bible had to do with real people and, presumably, real events.

In Drake's case, this enlightened approach backfired. "I saw that religious history . . . was an artificially narrow history because it ignored the many different ways of life and ways of belief there were in the world," he wrote in his 1992 memoir *Is Anyone Out There?* coauthored with Dava Sobel. If religious history was absurdly narrow, he reasoned, then why assume that all of human history was any less so? Why shouldn't there be many inhabited planets, each with its own history?

As far as Drake could tell, this question wasn't one that scientists thought a lot about. They hadn't talked or written much about it since the embarrassingly credulous days of the Martian canals. Science fiction writers wrote about aliens, of course, but he wasn't really interested in fanciful stories about imaginary aliens. He wanted the real thing. (He wasn't even remotely impressed at the time, and remains unimpressed to this day, by arguments that UFOs are alien spacecraft.)

Drake went on to college at Cornell, where he majored first in aeronautical engineering, then in electronics, and finally in engineering physics. He was still fascinated by the stars and was convinced that they must be home to other civilizations, but he also wanted to make a living when he graduated: The field of pure astronomy didn't seem to offer job security. He did take an introductory astronomy course and remembers being amazed by his first look at Jupiter through a telescope. But he also remembers that his professors never brought up the idea of extraterrestrial life. It was evidently too much of a fringe topic.

Then in 1951, during Drake's junior year, Otto Struve came to campus as a guest lecturer. This was the same Struve who had helped establish theoretically that planets could be common throughout the Milky Way. During one of his talks, Struve not only reemphasized that point but went on to make the radical suggestion that with so many planetary systems, it would be only reasonable to expect that many of them were inhabited. Almost a half-century later, sitting in his office at the SETI Institute, Drake still remembers Struve's words as electrifying. "This was a very astounding idea," he says, "especially because it came from a supposedly very conservative astronomer. It planted seeds in my mind."

For the first time Drake began to realize that extraterrestrial life might be a legitimate scientific question and not just his own personal daydream—at least one eminent astronomer had thought about it. He began to wonder how one might go about establishing the existence of extraterrestrials in a scientific way.

Partly in response to Struve's inspiration, Drake decided to go to graduate school for astronomy after a three-year stint in the Navy to pay off his ROTC scholarship. Thanks to his electronics background, he was trained as an electronics officer, and when he left the service to enter grad school at Harvard, his Navy experience made him a natural choice to work in the university's new radio astronomy laboratory.

Astronomers were just beginning to understand in the mid-1950s that the universe was fairly crackling with natural radio signals and that radio telescopes could provide an entirely new way of understanding the cosmos. Radio waves are a form of electromagnetic radiation, as are visible light, infrared, ultraviolet, and X rays. The differences are in wavelength. Visible light has a very short wavelength—the distance from peak to peak or trough to trough of these waves of energy is on the order of only a few hundred billionths of a meter. That means that about ten quadrillion peaks pass a given point every second: Light has a frequency of ten quadrillion cycles per second. Radio has frequencies of millions or thousands, and wavelengths of inches and feet. The numbers on a radio dial, which represent those frequencies, are colors just as red and green and blue are colors. It's just that our eyes can't see radio waves; our receivers are tuned to a different wave band.

But radio telescopes can see these waves, and the universe they perceive is completely different from the universe we see with our eyes and our optical telescopes. Jupiter outshines the Sun in radio wavelengths; the Moon barely shines at all. Nor do most stars. Interstellar gas clouds, by contrast, glow brightly with radio waves, even though they're almost invisible in ordinary light. It was reasonable to think that astronomers could learn all sorts of new things about the cosmos by looking at it through what were, in effect, different glasses. (As it turned out, they were right: A short list of the discoveries made by radio telescopes includes quasars, pulsars, and the cosmic afterglow of the Big Bang itself.)

Radio telescopes, like radios, depend on all sorts of complicated electronic instruments to process raw data and put it in a form that can be understood by humans—rectifiers, amplifiers, and other devices that most astronomers are clueless about. Not Frank Drake the Navy electronics officer, though, and so he became a radio astronomer.

It didn't take long for Drake to realize that these new telescopes might provide a direct connection to his longtime secret fascination. Humans used radio waves to communicate over long distances, so why wouldn't aliens? It was a rediscovery of Marconi's reasoning of forty years earlier, but the radio telescopes of the 1950s were millions of times more sensitive than Marconi's primitive receivers—partly because of their enormous size (the Harvard telescope had a dish-shaped receiving antenna sixty feet across) and partly because of their signal-boosting electronics. Marconi would have been lucky to hear a broadcast from Mars, but a modern radio telescope might be sensitive enough to pick up a broadcast from deep space.

Not that Drake ever tried. He wouldn't have dared. Then, as now, telescopes were in enough demand that they could only be used at approved times for approved research. Struve's example notwithstanding, looking for E.T.s was not something Drake would have had the nerve to propose. Even so, he almost went into heart failure one night when, during a routine observation of the Pleiades star cluster as part of his graduate dissertation, he picked up a regular, repeating signal. It was clearly artificial.

Looking for E.T.s was out of the question, but finding them by accident—well, who could complain about that? Drake was too good a scientist simply to run from the building screaming "eureka," however. First he had to rule out a more mundane explanation for what he seemed to have discovered. If he didn't and there really was such an explanation, somebody else would figure it out pretty fast, making Frank Drake look like an overeager, foolish kid. On the spot he invented one of the standard self-debunking tests that searchers for extraterrestrial intelligence still use. He moved the tele-

scope away from the Pleiades. The signal persisted. It was probably local radio interference.

Nevertheless, Drake was fired up. Yes, this was a false alarm, but it got
him to think more seriously about how one might go about detecting an
alien signal—what sort of assumptions one would have to make about the
signal itself and what sort of telescope one would need to look for it. He
made the conservative assumption that the aliens would be no more advanced than we were, and, therefore, whatever antenna was being used to
send the radio signal (a radio dish can send as well as receive) was no bigger
than the one he was using. By that standard, the Harvard dish wasn't big
enough. It couldn't pick up a signal from a comparable dish even as far
away as Alpha Centauri, the brightest member of the nearest star system to
our own.

So for several years all Drake did about it was think—through his graduation from Harvard and on into his first job, on the staff of the newly
founded National Radio Astronomy Observatory in Green Bank, West Virginia. The leading optical observatories were by now all out west, where the
sky was generally clear and where the mountains let you put up a telescope
above at least some of the Earth's roiling atmosphere. Neither factor mat-
tered for radio astronomy, though: Radio waves sail right through clouds,
and atmospheric turbulence doesn't bother them much. Green Bank, nes-
tled in a rural Appalachian valley, was an ideal location—reasonably close
to the population centers of the eastern seaboard but surrounded by hills
that would block out most interference from terrestrial broadcasts. Radio
and television reception in the Deer Creek Valley is poor, which suits the
astronomers just fine.

It was here that Frank Drake finally got up the nerve to search for aliens.
Green Bank had an eighty-five-foot dish—large enough, conceivably, to pick
up a similar-sized dish on the score of stars that lie within twelve light-years
of Earth. Drake realized that in order to detect what he thought were the
likely frequencies of alien broadcasters, he'd have to build a new kind of receiver to attach to the dish. He deliberately chose to build one that would be
useful for addressing a different, far more mainstream scientific problem.
Theorists had suggested that if magnetic fields existed in deep space—the

same sort of fields that make compasses work on Earth—then they should have a measurable effect on the radio hum emanating from interstellar clouds of hydrogen gas. Drake built his receiver so that it would be capable of detecting this so-called Zeeman effect; nobody could accuse him of wasting the taxpayers' money building an instrument for such a foolish purpose as pure alien-chasing.

Still with some nervousness, Drake brought up the idea to the lab director, Lloyd Berkner, and Berkner loved it. Project Ozma, named for the princess of L. Frank Baum's land of Oz, was now official. Drake went to work, deliberately keeping the project a secret from other astronomers and from the press. He could just imagine the silly headlines if someone found out what he was doing.

He was evidently not the only one, however, who realized that radio waves were the way to look for extraterrestrials. Even as he was putting his detector together in the fall of 1959, an article entitled "Searching for Interstellar Communications," written by Philip Morrison and Giuseppe Cocconi, appeared in the prestigious British scientific journal *Nature*. Morrison and Cocconi were proposing precisely what Drake was doing, except that these already eminent scientists were better positioned to risk taking scientific heat for talking about it publicly.

Nearly forty years after the *Nature* paper appeared, Morrison still remembers precisely how he and Cocconi had come up with the idea. Morrison is confined to a wheelchair, a result of a boyhood polio attack that has caused his muscles to atrophy. He spent World War II working on the Manhattan Project; the plutonium core for the very first atomic bomb rode out to the Trinity test site in New Mexico on Morrison's lap in the backseat of a car. He later went on to a distinguished career as a physicist at Cornell and M.I.T., as an antinuclear activist, and as a talented popularizer of science. The award-winning film *Powers of Ten*, which graphically shows the relative sizes of things in the universe all the way from subatomic particles to superclusters of galaxies, was written by Morrison and his wife, Phyllis.

Philip Morrison was still at Cornell at the time he and Cocconi wrote their paper. He'd been there a few years earlier when Frank Drake was an undergraduate, but the two never met and certainly never discussed SETI.

At Cornell, Morrison was caught up in the excitement over the university's new cyclotron, a much smaller ancestor of today's giant particle accelerators. Among other things, the cyclotron spewed out gamma rays—yet another form of light but with an even higher frequency and shorter wavelength than X rays. "None of us," said Morrison with the relish of a wine connoisseur recalling the first time he'd tasted a particularly fine vintage, "had ever seen gamma rays of that intensity, control, and energy."

That got Morrison thinking about the properties of gamma rays in general. They would be able, he realized, to travel long distances in space without distortion. You could do gamma ray astronomy just as you do radio or optical astronomy. He estimated it would take maybe five years to build the necessary equipment; in fact, it took three times as long. "We had," he admits, "the usual theorists' naïveté about how hard it is to do experiments."

One day Morrison's colleague Giuseppe Cocconi came to his office and raised an interesting question: If they could somehow aim their accelerator at the sky, couldn't they send gamma rays across the Milky Way to be received by alien astronomers if they existed? And mightn't aliens have come up with the same notion? Morrison's response: "It's a fine idea, Giuseppe, but should we not be a little more thorough? Just because gamma rays can be seen from across the Milky Way, that doesn't make them the best medium for communicating."

So Morrison and Cocconi spent the next few months looking into all the wavelengths of electromagnetic radiation, including radio, infrared, visible light, ultraviolet, X rays, and gamma rays. They finally decided that radio waves would be best after all. They are simple to generate and simple to detect—for us and also for aliens, they presumed. The radio band includes lots of frequencies, though, so you have to figure out what channel the aliens might be listening to and what channel should be used to listen for them.

Morrison and Cocconi again found an answer. The universe is made up mostly of hydrogen. When hydrogen atoms are disturbed, they emit radio waves with a frequency of 1,420 million cycles per second and a wavelength of twenty-one centimeters. Radio astronomers throughout the universe would naturally build receivers sensitive to this most common frequency and would spend much of their time studying it. So if they were interested in signaling their existence, they'd undoubtedly do so close to that frequency where scientists would most likely be listening. "We therefore feel that a discriminating search for signals deserves a considerable effort," wrote Cocconi and Morrison. "The probability of success is difficult to estimate, but if we never search, the chance of success is zero."

By the time the paper appeared in print, Morrison was out of the country. "I was doing some traveling," he says, "and as I went from place to place, I kept hearing from journalists. They had evidently heard about the paper and checked with other astronomers and physicists, who assured them that this idea was perfectly respectable. I met with reporters in Italy, I believe, Israel, India, Japan, and finally back home in the eastern United States. And there I learned to my astonishment and delight that a radio astronomer named Frank Drake had independently prepared an apparatus for doing precisely what we'd suggested." Morrison heard about Drake's work because Otto Struve, who had now taken over the directorship of the National Radio Astronomy Observatory from Lloyd Berkner, was worried that Drake and NRAO wouldn't get credit for the idea. He overruled Drake's vow of secrecy and announced the project at a lecture at M.I.T., emphasizing that it had already been in the works for a year.

Finally, on August 8, 1960, Project Ozma was ready to go. Drake and his two assistants aimed the eighty-five-foot dish at his first target star, Tau Ceti, a Sun-like star about twelve light-years from Earth in the constellation Cetus. He tracked it for a few hours, then turned the dish toward his second target, Epsilon Eridani, a little farther out. Suddenly the pen on the team's chart recorder started squiggling wildly, and the audio output began to screech. Just as he had with the Pleiades five years earlier, he moved the telescope away from the target. This time the signal stopped. When he moved the dish back, though, the signal didn't start up again. It was gone.

For five days Drake looked at Epsilon Eridani, hoping that the mystery would reappear. He also took the precaution of aiming a second, small antenna out the window in a completely different direction. Maybe the signal had really been terrestrial interference that coincidentally shut off when he moved the big dish. This way he'd already have a receiver pointed away if the main receiver picked up the signal again.

That's just what happened. Epsilon Eridani began broadcasting again, but the window antenna heard the same signal at the same time. It was interference after all. In any case, Project Ozma was supposed to last only two weeks. At the end of that time Drake packed it in and went back to looking for the Zeeman effect (which he found, as it turned out, so his receiver wasn't built in vain). The experiment was over.

But Project Ozma wasn't soon forgotten. Although he didn't find any alien radio stations, the fact that Drake had taken the trouble to look created a stir. Green Bank, where the observatory was located, is in the middle of nowhere, but reporters began showing up to see the apparatus and talk to Drake.

Then one day, about a year later, Drake received a telephone call from J. Peter Pearman of the National Academy of Science's Space Science Board. The board had created a committee on exobiology, and it had taken up the still open question of whether life might exist elsewhere in the solar system—not the canal-building Martians of Percival Lowell but the microbes that serious scientists thought might be on the planet. NASA had started planning a series of probes to be launched to the planets, and the search for Martian microorganisms, alive or dead, was an important part of the plan. Pearman believed that Drake's work was closely related to this new science of exobiology and should be brought into the fold. He asked Drake to organize a conference at NRAO where experts in all aspects of the search for life could meet and exchange ideas.

Drake agreed, after making sure Struve was amenable, and between them Drake and Pearman put together a guest list. Struve was at the top, thanks both to his position and to his pivotal role in establishing the likelihood of planets in the Milky Way for E.T.s to live on. Cocconi and Morrison were included, too, and so was Carl Sagan, a young astronomer whose official field of expertise was planetary atmospheres but whose powerful interest in the question of life on other worlds had gotten him a spot on the exobiology committee.

Drake also invited Bernard Oliver, vice president for research at Hewlett-Packard, who had become extremely interested in SETI; Dana Atchley, an electronics expert who had donated a sophisticated radio amplifier to Project Ozma; Melvin Calvin, whose experiments a decade earlier had suggested how organic molecules might have arisen on the young Earth (Calvin would win the Nobel Prize during the conference for his research into photosynthesis); Su-Shu Huang, an astronomer who had thought about what kinds of stars might be most hospitable to life; and, finally, John Lilly, whose work in trying to communicate with dolphins could—so he claimed—help scientists understand what it would be like to communicate with intelligent aliens.

The invitees were contacted and accepted, to Drake's relief. This group of eminent scientists didn't think the subject was flaky. Now all he needed was an agenda. He began thinking, as others had before him, about all the factors that would have to come together to create a civilization detectable across interstellar space. You'd need the right sort of star, the right sort of planets, the right conditions for life to arise, and so on. Unlike anyone before him, though, Drake set down each of these factors in mathematical shorthand. That made sense. If you wanted to estimate how many civilizations there were in the galaxy, you had to work out the probability of each factor and then multiply them together to get the answer.

When he was finished thinking, he had a mathematical equation, and when the conference formally convened a few days later, he stood up and wrote it on the blackboard: $N = R^* F_p N_e F_l F_i F_c L$. Of all these factors, the conference participants had reliable information only on the first: Sun-like stars are born, on average, about once a year. That didn't stop them from speculating about the rest, though. Any competent scientist can come up with a plausible estimate on just about anything, given even the tiniest bit of actual knowledge. As long as it's presented honestly and intelligently, such an estimate is valid as a discussion point at least. The fact that Philip Morrison was well read on many subjects didn't hurt. He had plenty to say not only about astrophysics but also about such things as convergent evolution and the forces that drive civilizations.

Over the next few days the scientists went systematically through the terms of this useful new equation. The number of stars that form planets, they decided—even after you allow for the fact that more than half the Sun-like stars in the Milky Way are part of multiple-star systems and thus, presumably, not amenable to forming planets—was perhaps 20 percent. The number of planets hospitable to life was between one and five per solar system. The number of such planets where life could arise was all of them, said both Calvin and Sagan, with enough intensity to convince the others. The number of these where intelligence would arise was, again, all of them. The number of these that would develop communication was 10 percent to 20 percent, they decided. And, finally, the length of time that the average civilization would continue was somewhere between one thousand and one million years—admittedly a liberal range.

As it turned out, most of the factors canceled each other out. The number five over here got multiplied by the number one-fifth over there, and so on. They all multiplied out to one—leaving the final factor sitting by itself. Mathematically, the only number that seemed to matter (assuming the other estimates were more or less accurate) was L. If the average civilization does in fact endure for between one thousand and one million years, then the number of communicating civilizations in the galaxy is between one thousand and one million.

If so, it's worth looking for them, agreed the members of the Order of the Dolphin—for that's what the Green Bank conferees named themselves at the end of the meeting, in reference to John Lilly's work. (Melvin Calvin later sent all the participants gold dolphin-shaped lapel pins.) NASA, flush with money and feverishly planning its exploration and conquest of the solar system, agreed. The space agency began putting small amounts of research money into Drake's and Morrison's and Cocconi's research, which complemented its plans to look for life on Mars. While most scientists would continue to nibble away at the newly minted Drake equation from its left-hand side, trying to understand the nature of planets and determine whether life is a common event, SETI researchers came at the equation from the right. After several thousand years of speculation, one simple SETI detection would prove immediately and unambiguously that at least two intelligent races share the Milky Way, with the implication that there are probably many, many more.

Still, SETI was a long shot. Unless someone got unexpectedly lucky, the only way scientists would find the answer to the question of whether life existed on other worlds was slowly, one careful step at a time. And the first step, everyone agreed, was to prove that alien life had a place to live—to demonstrate, finally, that planets really do orbit stars beyond the Sun.

Chapter 3

PLANET QUEST

If he had grown up in a different kind of family, lived in a different neighborhood, gone through a different school system, majored in a different subject in college, gone to a different graduate school, or even had a different personality, Paul Butler would never have become part of the most successful planet-hunting team in history. Just about everything in his life pushed him toward that role, but only indirectly.

His family, for example, wasn't especially scholarly or science-friendly. Quite the opposite, in fact. His father was a Los Angeles policeman. Butler was raised in a working-class neighborhood with a lot of people who would now be Rush Limbaugh fans. When he was about twelve, he decided his father was wrong about almost everything. This is not an unusual occurrence, of course, and what generally follows is adolescent rebellion—harsh music, disreputable friends, strange hair.

In Paul Butler's case, the rebellion was intellectual. Around that time he discovered a book in the school library: Bertrand Russell's *Why I Am Not a Christian*. It came as a revelation to him to learn that science could provide a basis for living that was every bit as profound as religion. "It was cool," he recalls, "to read about people who were willing to be arrested or even executed for saying incredibly rational things."

The word "cool" is an essential part of Butler's vocabulary. He may be a world-renowned scientist, but he talks like the quintessential California guy. He doesn't exactly look the part, however. Now in his early thirties, Butler is over six feet tall and big—he tends toward overweight when he can't make it to the gym regularly. He is prematurely bald and has a closely cropped beard. He invariably looks rumpled, even when he hasn't been up late the night before finding new planets. His shirttails do not behave themselves.

But while his speech is relaxed, even laid-back, his words convey a great deal of passion. It doesn't take long to figure out that Butler's world is pretty much divided between heroes and villains. Among the latter are Berkeley street criminals ("There is just this incredible amount of pond scum who are breaking windows and stealing stereos") and some astronomers ("He has real mental problems," Butler says of a well-known competitor. "It's a sort of Napoleonic complex. I don't think there's any hope for him"). Geoff Marcy is used to this sort of vehemence from his partner by now, but it does cause problems when Butler's private comments reach the subjects' ears. Marcy has had to mollify irate colleagues more than once.

It's equally easy to figure out who the heroes are. When Butler launched into his adolescent rebellion against his anti-intellectual background, most of the scientists he read about were astronomers, and most were active during the period in history when the notion of many worlds in the universe was just beginning to be accepted. One of his early heroes was Giordano Bruno. "They turned him into a martyr," says Butler. "That puts Bruno at the level of, say, Martin Luther King. A guy like Bruno blows away any scientist who's living today." Kepler was another hero. "He was brilliant, but beyond that, it took him ten years of intensive work before he managed to wrangle through his laws of planetary motion."

Yet while astronomy was his first love—he built his own telescope while in junior high and learned his way around the constellations—Butler graduated from college with a degree in chemistry. He started at Georgia Tech, spent some time at Harvey Mudd College in southern California, and wound up at San Francisco State University, where he graduated in the early 1980s.

In graduate school at San Francisco State, Butler began to concentrate on what really interested him: physics and astronomy. The first couple of years in grad school are like the last couple in college: You mostly take courses, though at a much higher level. It isn't until the third year that you start the transformation from student to working scientist. You find a professor to serve as a mentor, or the professor finds you, and you start to work, under his or her supervision, on a thesis—an actual unsolved scientific problem, or at least a piece of one. You're no longer going through lab exercises where everyone knows how it's going to come out. You're no longer dealing with hypothetical situations. Usually you end up working on the same problem the professor is working on; you provide an extra mind and a pair of hands, the professor provides guidance and on-the-job training.

You don't even have to discover something. Working scientists frequently perform experiments in which they discover nothing—no effect when you'd expect to see one. Finding nothing can be just as important as finding something. At the end of the nineteenth century, for example, when Albert Michelson and Edward Morley found no evidence of cosmic ether, the hypothetical substance that light waves presumably rippled through, their nondetection forced physicists into a radical rethinking that eventually led to Albert Einstein's special theory of relativity. But whether you find something or not, getting your Ph.D. depends on how thoroughly and thoughtfully you do the research, and thus on how convincing your conclusions are.

As it happened, Paul Butler was searching for a thesis topic just as Geoff Marcy was searching for a graduate student to help him. Marcy, too, had grown up in the Los Angeles area, in the San Fernando valley. "I'm an original valley boy," he says, although he speaks in the universal, non-regional accent many scientists have (its principal characteristic is precise pronunciation). You can't easily imagine him calling anyone "dude." You don't have to imagine it with Paul Butler.

Marcy's parents were well educated. His father was an aeronautical engineer who worked on the space shuttle, while his mother was an anthropologist. Marcy is about ten years older than Butler and about six inches shorter. He's slimmer, though he is starting to thicken around the middle. He generally looks tidy—shirttails tucked in, hair in place—and is perfectly at ease in a jacket and tie: Butler looks and feels as though he's wearing a slightly ridiculous costume. Marcy, whose hair is receding about as fast as Butler's is, wears a goatee, not a full beard. Marcy could be a headwaiter; Butler could be a football player in perpetual off-season. And unlike Paul Butler, Marcy is highly restrained in his speech. He is almost always polite when he talks about his competitors, even when he's saying something negative about them. Shoddy science or unfair criticism seems to sadden him, while it infuriates Butler.

Like Butler, Marcy was captivated by the heavens as an adolescent, although in his case it was thanks to his family, not in rebellion against all they stood for. At fourteen, he remembers, his parents bought him a small telescope, which he would wrestle out the window of his bedroom every night onto a flat roof so he could study the heavens. His favorite hobby was to look at Saturn and watch how one of its moons, Titan, changed position month after month. He charted the motions, calculated the period of Titan's orbit, and compared his results with what the textbooks said—usually favorably.

Unlike Butler, Marcy didn't take any significant educational detours: He went straight through U.C.L.A. with a double major in physics and astronomy, and then to graduate school at the University of California, Santa Cruz. His own thesis, researched and written under the supervision of an astronomer named Steve Vogt, involved trying to see whether Sun-like stars had magnetic fields, as the Sun does. Marcy concluded that they did, but some astronomers didn't find his work convincing, which led Marcy to doubt himself badly. "I didn't have any confidence in my own ideas," he told me. "I was in a depression. I was convinced I was an impostor, that I didn't belong with all of these high-powered people."

This was clearly not an objective view, since on graduation Marcy was awarded a prestigious postdoctoral fellowship—the paid apprenticeship most scientists go through before they become full-fledged faculty members—at the Carnegie Observatories in Pasadena, California. But if anything, being at

Carnegie made Marcy feel worse. He was surrounded by a group that was even more select than at Santa Cruz. He got up every morning feeling really bad. He was convinced he wasn't smart enough to be an astronomer.

Finally, one morning in 1983, Marcy dragged himself out of bed and into the shower as usual, but instead of turning off the water when he was finished, he just stood there thinking. He knew he had to get himself out of what had become a perpetual depression. "I'm not Einstein," he thought. "I'm never going to be. So what am I going to do—beat myself up over it for the rest of my life?" He recalled how as a kid he had had posters of the planets plastered on the walls of his bedroom and had stayed up half the night for the pure joy of exploring the universe through his telescope and then had sat glued to the television to watch humans take the first steps on the Moon in July 1969. If he could somehow reconnect with the sense of wonder he had felt back then, he might be able to get excited about astronomy again. "I have to find something to work on," he told himself, "that addresses a question I care about at a gut level." It also had to be something difficult to do. There wouldn't be much satisfaction or self-respect in solving an easy problem.

Then it came to him. One question that had always excited him was whether our solar system was unique or whether there were other inhabited planets in the universe. With the water still pouring down his back and with steam clouding the windows, Geoff Marcy decided to take on the second element of the Drake equation. Unless SETI researchers managed to snag an alien signal sometime soon and prove all in one shot that the answer was yes, Marcy would devote his career to moving the Drake equation forward a notch—something that hadn't happened since Frank Drake had written down the equation two decades earlier. Marcy decided he would look for planets.

Marcy knew that there were two ways to answer the question. You could spot a planet directly, orbiting around its parent star. That was impossible in the early 1980s, though, and it remains impossible today. At interstellar distances, a planet is so close to a star that their images tend to merge, just as an oncoming car's headlights seem to be a single light when seen from far away. You'd need a telescope that could focus sharply enough to separate planet from star, and even the Hubble couldn't do that. Beyond that, a planet was too dim for the most powerful telescope to pick out, even around the nearest stars. It would be like trying to observe a candle at a distance of five hundred miles, with the candle sitting next to a billion-candlepower searchlight aimed right at your face.

The other way to find a planet was indirectly, by looking at the influence of planets on stars. Physicists and astronomers know that there's no such thing in the universe as one object orbiting another; they orbit each other. When a moon goes around a planet or a planet goes around a star, what is really happening is that the two objects are orbiting each other around a common center of mass that lies somewhere between their centers. Even the Sun and the center of the Milky Way, thirty thousand light-years away, orbit each other.

If the objects are of equal mass, that center will lie exactly between them. You can't even say which is the orbiter and which the orbitee. If they are of very different masses, though, the center of mass will be much closer to the more massive object. It is as if a child and an elephant were on a seesaw; despite their vastly different weights, they would still be able to balance as long as the seesaw was extremely long and the elephant's end much closer than the child's end to the center of balance. The Earth and the Sun orbit a common center that is deep within the Sun itself. The common center of the Sun and Jupiter, the most massive planet in our solar system, is about at the Sun's surface. Over Jupiter's eleven-year orbital period, the Sun pirouettes delicately around that common center, moving a distance equal to its own width, first in one direction and then in the other.

Like all modern planet hunters, Geoff Marcy knew that this sort of motion was what he had to look for. Depending on its orientation, that motion would be either side to side or front to back—the former if you were looking down on a solar system from above, the latter if you were looking edge-on. It could also be a combination of both if you happened to be looking from some angle in between.

If it was side to side, you'd actually see the star move physically against the backdrop of the sky, and Marcy knew some astronomers were already looking for this effect. If it was front to back—what astronomers call radial velocity, since the radius of a circle stretches from the outer edge to the center—the star would stay fixed in place, but in principle you could still measure its motion.

You'd do this by spectroscopy, by taking the light coming from the star and smearing it out into the colors that make it up. Unless it's a laser, any light-emitting object emits a mix of colors of light, no matter how it appears to the eye. Raindrops reveal the true nature of sunlight by smearing its seemingly yellow-white color into its component rainbow. A prism or the optical surface known as a diffraction grating will do the same thing. And they'll do it as well to faint starlight gathered and focused by a telescope.

If you could capture a slice of rainbow and look at it under a microscope, you'd see that it isn't actually a perfect wash of color. It is striped with very fine dark lines. The reason is that while the sunlight shines with many colors, not all of that light makes it into space. The Sun has an atmosphere, just as the Earth does, and light has to shine out through this atmosphere. As it does, the molecules of gas that make up the solar atmosphere absorb a little bit of sunlight—each type of molecule absorbing light of a different specific frequency depending on the molecule's atomic structure. The light of those frequencies—the absorption frequencies of these gas molecules—is trapped. It never escapes. And because we perceive the frequencies of light as colors (just as we perceive sound waves of a given frequency as musical notes), the result is a rainbow of sunlight with missing colors. Most of the blue light, say, coming from the Sun still gets here, but a few very particular shades of blue—such thin slices of blue that they make no perceptible difference to the naked eye—don't.

These so-called absorption lines tell plenty about the Sun or about a star or even about the combined starlight of a distant galaxy. The colors that are missing correspond to specific molecules; they're a tip-off to what gases are present in those far-off objects. In fact, Joseph von Fraunhofer's discovery in the 1800s that the lines he saw in sunlight matched those he saw in the lab proved that the Sun is made of the same elements found on Earth.

But the lines say much more than what the Sun or a star is made of. Light—or, more formally, electromagnetic radiation—travels in waves, and in many ways it behaves like any other sort of wave. (It also behaves differently from every other sort, but that's beside the point here.) When you pack waves closer together, their frequency gets higher—the peaks and valleys of the waves come more often. When you stretch waves, the frequency gets lower. If it is a sound wave, higher frequency means a higher note—which is why the motion of an approaching police car, squeezing the sound waves in front of it, makes a siren sound higher and why the sound drops dramatically in pitch as the car passes by, as the sound waves stretch to overcome the increasing distance.

Precisely the same thing happens with light waves, except that with light, a shift toward the higher-frequency, blue end of the rainbow is the equivalent of a higher pitch, and a shift toward the red end is lower. It is nearly impossible to notice this effect on the overall color of light coming from a star, but the absorption lines are a dead giveaway. A trained observer knows the patterns and can identify individual gases. They are like the markings on a ruler—they're always in the same positions, and thus it's easy to gauge when the ruler has shifted even a tiny bit. If an astronomer looks into a spectroscope—a device that turns white starlight into colorful starbows—he or she can judge whether the light is red-shifted—moving away from us—or blueshifted as it approaches.

This is how astronomers can tell that the universe is expanding. Absorption lines in the light from almost every galaxy they can see are red-shifted—the galaxy is moving away from us. That is also how, in theory, observers might detect the back-and-forth of a star being tugged on by a planet. It was the way Geoff Marcy figured he'd go about it. Spectroscopy, the analysis of starbows, was his specialty. His graduate work on magnetic fields had involved looking for the subtle effects that those fields had on absorption lines. Marcy knew that Doppler measurements could work in principle. (Motion-induced red and blue shifts are known collectively as Doppler shifts, after Christian Doppler, an Austrian physicist who described the effect in 1842.) Marcy had no idea, however, standing there in the shower, whether he could actually make such delicate measurements.

Galaxies are flying apart at hundreds of thousands, even millions of miles per hour—a huge effect that is easy to measure. But the motions of stars under the influence of planets is minuscule. To wobble by its own width in the course of Jupiter's eleven-year orbit means that the Sun has to go about a million miles in one direction over about five and a half years, and then a million miles in the opposite direction in the same time—an average speed that works out to about twenty-eight miles per hour. If Jupiter were more massive, the wobble and thus the speed would be greater; the same would be true if it were closer to the Sun. But not by much. In any case, everybody figured that our solar system was typical—an assumption based on essentially no information.

Marcy soon realized just how tough a problem he had taken on. It turned out that most astronomers would congratulate themselves if they could measure a Doppler shift to within 1 kilometer per second, or about 2,200 miles per hour—a tiny error when you're talking about galaxies moving apart at 100,000 miles per hour. But the same error is impossibly large when you're measuring a planet-induced wobble of 20 or 30 miles per hour.

Clearly, Marcy would have to figure out how to slash his measurement errors if he was going to find planets. And although he had no idea how, he swore he was going to do it. Luckily, his Carnegie fellowship gave Marcy an extraordinary amount of freedom and lots of telescope time. After months of concentrating on the problem, he learned that by being extraordinarily meticulous he could reduce his measurement errors to plus or minus two hundred meters per second. He was only about 1,000 percent off—incredibly good but still not good enough.

The reason Marcy couldn't do better was that he was measuring spectra the way everyone else did. In order to know how far a given absorption line is from where it should be, you need something to compare it with—a non-shifted version of the same line. What astronomers usually do is rely on a reference lamp, a light bulb filled with gas whose own spectrum is thoroughly understood. You shine the lamp into the spectrometer, take note of where the lines fall, then find the same lines, red or blue shifted, in starlight and measure the difference.

That is much too crude for planet-hunting, though. In the few seconds it takes to move the telescope from star to reference light and back again, the whole system can go out of adjustment by enough to throw the measurements off. One solution might be to use mirrors and direct the lamplight into the telescope without moving it. But even then the light will enter the telescope at a different angle from the starlight and bounce around inside in a subtly but fatally different way.

What you really need is a set of reference lines that enter the telescope at *exactly* the same angle as the starlight. In 1973, long before Geoff Marcy had embarked on his project, astronomers had realized that such a set of lines existed. The Earth's atmosphere is made of gases whose molecules absorb specific frequencies of light, just as the molecules in a star's atmosphere do. When starlight passes through the atmosphere and into the telescope, it should carry the imprint of these atmospheric reference lines along with its own intrinsic spectral lines. And since the atmosphere obviously isn't moving toward or away from Earth, atmospheric lines should stay put while the star's lines are shifted in one direction or another.

Unfortunately, the atmosphere's reference lines fall mostly in the infrared part of the light spectrum, a region beyond human vision, while most of the lines in starlight are in the visible range. The two sets of lines don't overlap much, and you can't easily see them with the same detectors. Besides, the atmosphere is not as stationary as you might think. Updrafts and downdrafts send pockets of air rushing toward and away from the telescope at scores of miles per hour. That red-shifts and blue-shifts them just enough to garble the signal from an orbiting planet.

In the late 1970s, someone else had thought of a way around this problem. Bruce Campbell was a postdoctoral student in astronomy at the University of British Columbia. Campbell and his colleagues realized that they could solve the problems of a moving atmosphere by creating an artificial atmosphere of their own—by finding a gas with nice, prominent absorption lines, trapping it in a bottle, and putting the bottle in front of their spectrometer. Starlight would arrive from many light-years away, bounce around in the telescope, and at the very last minute pass through the gas in the bottle, taking the imprint of the gas's lines into the spectrometer. The gas would be the very last thing starlight would encounter before entering its ultimate resting place.

It worked. Where conventional measuring techniques had been getting errors of one thousand meters per second—and Geoff Marcy was squeezing that down to two hundred—Campbell and his colleagues had managed to drive their errors all the way down to twelve or thirteen meters per second. You couldn't quite detect Jupiter at that level, since the planet itself would generate a signal no greater than the measurement errors, but you could find oversized planets and smaller planets closer in. Unfortunately, the gas Campbell used was hydrogen fluoride—terribly corrosive, explosive, and so toxic that it can soak right through your skin and kill you.

Marcy read about the Canadians' technique, realized right away how ingenious it was, and decided to copy it. He didn't like the idea of hydrogen fluoride, though. This was due in part to its nastiness and to the elaborate safety procedures he'd have to use if he worked with it. It was also due to the chemical's inherent limitations. Hydrogen fluoride has prominent spectral absorption lines, but they cover only a small part of the visible light spectrum and thus a small physical space in the spectrometer. If the spectrometer is less than perfectly uniform in shape—for example, if the size of its pixels, or light-sensing elements, are slightly different at one edge of the light detector than at the other—then a calibration at one end won't accurately gauge the star's overall red shift. And if the star is very faint and the lines that lie close to hydrogen fluoride's lines happen to be too dim to pick out, you won't be able to calibrate the star's red shift at all.

Marcy needed a better molecule. For that he needed a chemist, one who had a thorough understanding of what he wanted to do, who was also knowledgeable about astronomy, not just an expert who could fill a prescription.

By now Marcy had left the Carnegie Observatories and had taken a faculty position at San Francisco State University. Paul Butler had just graduated from San Francisco State with a degree in chemistry and a powerful desire to get back into astronomy. Marcy approached Butler, explained the project and the problems with hydrogen fluoride, and said he needed someone to help him figure out a better way.

This suited Butler just fine. "I *really* hate to do standard sorts of stuff," he says. He and Marcy were ideally suited to each other: They both thrived on doing difficult projects that were off the beaten path but that had potentially big payoffs. "It wouldn't be any fun," says Butler, "to do what everyone else is doing."

For the better part of a year Butler hung out with chemists and those in charge of the chemistry stockroom. He found a promising molecule called fluorine dioxide, which is highly explosive. It ultimately proved too unstable, and Butler went on to test and discard dozens of other chemicals. He finally settled on iodine. Unlike hydrogen fluoride, iodine gas is nearly harmless; you can handle it and keep it in ordinary glass, and it has lines all across the visible spectrum, from the lowest-frequency red to the highest-frequency violet. There is just one complication: Iodine has what Butler calls "an incredibly, brutally, messy, ugly spectrum." It has so many lines that when you look at them through most spectrometers, they seem to overlap.

Still, there was no other satisfactory alternative to iodine. Butler reported his good news and bad news to Marcy, and they decided to go ahead. They had no idea how to use iodine; the spectrum of iodine was so complicated that they had no way of analyzing their data. But they began observing stars one after another and stored their observations on tape, saving them for the day when they could actually do something with them.

The telescope they used for most of these observations is the 120-inch reflector at the University of California's Lick Observatory on Mount Hamilton near San Jose. Mount Hamilton was first colonized by astronomers in the 1870s, thanks to James Lick, a land speculator who got rich during the California gold rush. In his old age Lick decided to erect a monument to himself, so he endowed the University of California with enough money to build the world's biggest telescope.

The observatory was completed about ten years after he died and was equipped with what was then the largest telescope in the world, a thirty-six-inch refractor—that is, a telescope that magnifies light with a lens rather than with a curved mirror. A few years afterward, the builders—Alvan Clark and Sons, the most celebrated telescope makers of the nineteenth century—constructed a forty-inch for Yerkes Observatory in Wisconsin, and that was it. Nobody has ever built a refractor bigger than that. A lens more than three feet across is so massive and yet requires such clarity and precision that it is almost impossible to manufacture.

The thirty-six-inch is still on Mount Hamilton, inside a dome that has inlaid wood floors and wood-paneled walls; it is a huge metal tube more than a yard across and fifty feet long, painted battleship gray. It's mounted on a concrete pier that rises twenty feet above the floor. The telescope is so long that it would gouge the floor if aimed straight up except that the floor can be lowered hydraulically. The motion doesn't affect the telescope because the pier that supports it extends down through the floor, without actually touching it, and right into the mountain. Even a century ago, astronomers were worried about vibrations getting into the telescope, so the pier is completely independent of the rest of the building. It is inside the structure but not part of it.

Refracting telescopes are always much longer than reflectors of an equivalent light-gathering power. A refractor gathers and focuses light through a series of lenses arranged in a line. A reflector does it all with mirrors. Light comes streaming into the telescope and bounces off a concave mirror, right back up the telescope. Because the mirror is curved, the light converges on a single point of focus. Reflectors are shorter than refractors because they use the same length of tubing twice. Sir Isaac Newton made the first working model in 1668, but it wasn't until the end of the nineteenth century that reflectors finally overtook refractors in quality.

As late as the 1920s, though, the Lick refractor was still among the most popular telescopes around. It was one of the first to incorporate the brand-new technology of photography. Astronomers at Lick were some of the first to routinely categorize stars by their compositions. They also set the standard in the 1920s for radial velocity measurements, the precursors for what Butler and Marcy do today.

Although the thirty-six-inch is no longer used for serious astronomy, it's still one of the best on Earth for stargazing. No one actually looks through large research telescopes anymore. The human eye has long since been eclipsed, by photography and, in the last two decades, electronic detectors, both of which are much more efficient at light-gathering. "Geoff caught the very tail end of the photographic era when he was a grad student at Santa Cruz back in the seventies," says Butler. "I just missed it—and I'm tickled pink about that. I don't need to be one of those guys who tells young astronomers how he walked through thirty miles of snow every day to coax data off a photographic plate."

Modern designers spend a lot of time figuring out where and how and
what sort of detectors they'll attach to their telescopes; they don't waste time
making it possible to look through them. The thirty-six-inch was built in the
days when the only way an astronomer could record an observation was on a
pad of drawing paper.

Astronomy professors from the University of California still bring undergraduate students up the mountain to gaze through the thirty-six-inch's eyepiece. If there's time, the students are led to a metal staircase that leads
down beneath the floor and into a sort of crawlspace that extends under the
wooden floor. Sitting in front of the concrete telescope pier, which is visible
here, too, as it descends through the building on its way into the mountain
itself, is a small basket of plastic flowers, marking the gravesite of James Lick.
His will stipulated that he be part of the telescope, so when they poured the
concrete about ten years after he died, they dug him up and put him inside.

The Lick Observatory complex is a lot more elaborate today than it was at
the turn of the century. It has modern dormitories, a small dining hall, and
even a small but fully accredited public elementary school, so the engineers
and telescope operators who keep the place running don't have to live elsewhere and commute. In the late 1950s the university installed a 120-inch reflecting telescope here. At the time, it was the second largest in the world,
after the 200-inch Hale Telescope at Palomar Observatory, 450 miles to the
south. "For a while there, it was kind of wild," says Butler. "In the sixties Lick
had the second-largest refractor in the world and the second-largest reflector,
the 120-inch. So it was the second-greatest observatory in the world, you
might argue."

By now the 120-inch is more like the twentieth most powerful in the
world, and it will lose several more notches in the next few years as telescope
designers complete a new generation of instruments. A smaller telescope has
less light-gathering power than a large one; for some projects the 120-inch is
as obsolete as the 36-inch refractor.

But for Marcy and Butler's project, it was perfectly adequate to start with.
They didn't need to gather a huge amount of light because they were looking for planets around the brightest, nearest stars. They didn't need a state-
of-the-art telescope.

What they did need was an extraordinarily powerful spectrograph. (The terms spectroscope, spectrometer, and spectrograph are used interchangeably, but they're not identical. A spectroscope is a device for smearing light out into its colors. A spectrometer incorporates a spectroscope but magnifies the image so the details of a spectrum can be seen more easily. A spectrograph records spectra for later study. Astronomers almost always use a single device that combines all three functions; confusingly, it has no name of its own, but is usually called a spectrometer or a spectrograph.)

Fortunately for Marcy and Butler, the one attached to the 120-inch is very powerful. It is known as a high-resolution spectrograph, which means it's especially designed to bring out a spectrum's finest details—especially the precise positions of spectral lines, making it possible to measure exactly how far they are out of position. The tradeoff is that in stretching light to its limits in order to wring out the details, you can spread it too thin to be detectable.

The Hamilton Spectrograph, as the Lick device is known, optimized for detail, can only get a useful spectrum out of a bright object like a star. "The galaxy guys can't even use it," says Butler. As a result, the spectrograph, which cost about $500,000 to build, was only in use about half the month. So the observatory knocked a hole in the side of the building, attached a small room outside, and put a twenty-four-inch telescope in it to feed light to the spectrograph when the big telescope was otherwise occupied. Whenever he could, Butler would drive up the mountain and sneak in some unofficial work on this small telescope—"guerrilla observing time," he called it. About half of Marcy and Butler's observations have actually been made with the twenty-four-inch over the years. Half the credit for any discoveries, Butler reasons, should go to the small telescope.

But all the credit should more fairly go to the Hamilton Spectrograph, and credit for the spectrograph in turn belongs to Steve Vogt, Geoff Marcy's thesis advisor from Santa Cruz. Vogt is a member of a small sub-fraternity of astronomers who, in addition to observing the heavens, actually design and build the instruments they use, in the tradition of Galileo and Newton. He is with Giordano Bruno in Butler's pantheon of astronomical heroes. He is, insists Paul Butler, the greatest spectrometer builder in history. "When we detect planets," he says, "we're sailing on Steve's ship."

Steve Vogt's ship sits in the basement of the 120-inch dome. To get to it, you have to walk past the huge, squat telescope with its ten-foot-diameter mirror and down a stairway that leads to a subfloor. No entombed benefactors down here—just enormous tangles of wires and concrete posts and stairways leading to various mysterious rooms. One such stairway has a motorized shoeshine brush at its foot, the kind found at the entrances to men's rooms in fancy hotels. It's a dust-control measure; the optical surfaces in the spectrometer are too delicate to be cleaned, so the astronomers do whatever they can to avoid getting the optics dirty.

Once your shoes are clean, you go up the stairway and into what looks like a large stainless-steel box. It's actually an antechamber for the spectrometer. Before you go into the spectrometer room itself, you have to walk over a sheet of sticky plastic, like a huge strip of Scotch tape that is adhesive side up. It makes a horrible ripping sound, but it gets the last bits of dust off the bottoms of shoes.

Finally, you step through the final door into what Butler calls the inner sanctum. You are literally inside the spectrograph. The room is perhaps six and a half feet high, fifteen feet wide, and fifteen feet deep. The whole room is on a slant. The floor resembles the seats of a stadium: three or four big concrete steps, going down. The ceiling slopes down at the same angle. The whole thing was clearly squeezed into an unused—and nearly unusable—corner of the building.

When the astronomers are taking data, this room is utterly dark. Light that has been gathered and focused by the telescope is bounced through a tiny window in the wall at the upper end. It then bounces off a diffraction grating—an optical surface that smears light into a rainbow in a way that is slightly different from that of a prism. Then it shoots across the room several times, caroming off various mirrors until it zips neatly into the final detector like a billiard ball into its designated pocket. A perfect shot every time.

The light detector at the heart of the spectrograph works best when it is cooled far below zero, to minimize electronic noise that could be mistaken for a wobbling planet. Before every night's observing run begins, therefore, one of the astronomers has to top off the detector with liquid nitrogen, which is stored in a large insulated stainless-steel vessel. It's like pouring iced tea from a thermos into a pitcher, except that this tea is at minus 196 degrees Centigrade.

So far all of this is standard for observing with the Hamilton Spectrograph. Before they do their specialized work, though, Marcy and Butler do one more thing: They step outside the room, go to the window that admits light from the telescope into the spectrometer, and move a cylinder about eight inches long and four inches across, its ends made of clear glass, into the path of the light.

This cylinder, filled with iodine gas, is the product of Butler's year in the chemistry lab nine years earlier; it makes Marcy and Butler's planet-detection system more sensitive than any other on Earth. When the cylinder swings into the light path, the iodine inside superimposes its own spectrum on the spectrum of starlight, a set of reference lines against which the astronomers can measure any wobbles a planet might be imposing on its star.

It can do this in principle, that is. In practice, despite Marcy and Butler's cleverness at merging their reference spectral lines right in with starlight, they were not able to make a decent measurement for the first six years or so of the project. The problem was that they were trying to detect an almost imperceptible effect. As Steve Vogt describes it, "If you took a ruler a couple of inches long and stood it on end, the amount it would shrink due to gravity is the kind of effect you're talking about. If you pick it up, the expansion due to heat from your hand is *one hundred times* the effect you're looking for. And you have to measure that."

Another way to look at the problem is in reference to the detector itself. In essence, the spectrograph is an extremely sensitive electronic camera, and like any camera, it records images as a series of dots. The dots on conventional photo paper are minuscule particles of silver halide, which turn darker as they're exposed to more light. The dots on an electronic detector—known properly as a charge-coupled device (CCD)—are tiny buckets of electric charge which trap electrons that are knocked loose by incoming light. If you get close enough to a photo, you can see the grain, and the same goes for a CCD image. Each grain in the latter is known as a pixel, for picture element (the dots on your computer or TV screen are also known as pixels), and they're a measure of how fine a level of detail you can record or display.

Vogt's Hamilton Spectrograph has unusually high resolution; it can record finer detail than most of the competition. But the effect of a Jupiter-size planet pulling back and forth on a star is so slight that it will shift the star's spectrum by approximately one one-thousandth of a pixel. Imagine the picture on your TV screen suddenly shifted by one one-thousandth of the width of one of the dots you can't even see unless you press your face against the screen. Not highly noticeable.

On the other hand, it's not impossible to measure, given a bit of cleverness. A CCD allows you to measure precisely how many electrons have fallen into each tiny electronic bucket. So if you have, say, 1,000 electrons in this bucket and zero in the next, and a while later you have 999 in the first bucket and 1 in the second, this means the whole image has shifted by one-thousandth of a pixel (the actual analysis is a little more complicated than this, you may be sure).

Once they've set up the spectrometer, Marcy and Butler go through an almost ritual series of steps in the control room of the 120-inch telescope, located just off the main dome. They go over the list of stars to be observed that night, make sure the software is properly loaded, and run the spectrograph through some electronic tests. They discuss the night's work with the professional telescope operator who is assigned to work with them that evening. (Astronomers rarely run big telescopes themselves anymore. These complex computer-controlled machines require more attention than a scientist is willing to divert from his or her own work.) They also go through the age-old astronomical ritual of checking the weather to see if any clouds are likely to spoil their work. Astronomers talk about the weather at least as much as farmers do. Near the California coast, where the Lick is, conditions aren't always ideal, especially in winter when storm systems blow in from the Pacific one after another.

Even with a fine instrument like the Hamilton, Marcy and Butler discovered, they still couldn't come close to making the kind of precise measurements they needed. The problem, which they realized only after some time, was that even with all its ability to see fine detail, the Hamilton Spectrograph had its limits. Its spectra were sharp but not razor-sharp. Aberrations in the optics would smear out the lines so that each line would cover several pixels at once. Moreover, the smearing was not symmetrical. There was more smearing to the right than to the left, and it didn't remain constant over time and wasn't consistent over the whole spectrograph. It was like taking a photo that was not just blurred but blurred by different amounts in different parts of the photograph. Worst of all, you knew that the photograph would be blurred in an entirely different way the next time.

What Marcy and Butler had to do was find a way to unblur the Hamilton's images. They finally did, but their solution was so complicated that it seemed insane.

What Marcy and Butler did was take their iodine cell to Kitt Peak National Observatory in Arizona, where the highest-precision spectrograph in the world resides. The Kitt Peak device is much more precise than the Hamilton because it was designed to take spectra of the Sun. It can afford to smear light far out without losing intensity, making it much more accurate than a starlight spectrograph could ever hope to be. The spectrum Marcy and Butler got at Kitt Peak is so perfect that Butler calls it "God's own spectrum." It's the true spectrum inherent in the iodine, seen with clear vision.

Armed with both God's spectrum and the Hamilton's, Marcy and Butler could then play a detective game to figure out exactly how the two were related. They created a computer program that takes the perfect spectrum and blurs it into exactly the same distorted spectrum they observed on the Hamilton. The program captures all the flaws in the Hamilton's optics in the form of a mathematical model, a horrendously complex series of equations. Then the equations can be run in reverse, taking any messy spectrum churned out by the Hamilton and transforming it into God's spectrum—of their cylindrical iodine cell or of any star observed. The equations recreate what God would see, for both star and iodine, and Marcy and Butler then compare the two to measure a star's Doppler shift.

But even then the precision is still not good enough for the very exacting measurements that Marcy and Butler need to make. So in one final step the astronomers break each observation into seven hundred parts, each representing a small slice of the star's overall spectrum just forty pixels wide.

For each chunk of spectrum they calculate the Doppler shift, and thereby the star's radial velocity, and then average them. It's this last crunching of numbers that gives them an accuracy of one one-thousandth of a pixel.

Writing a computer program that would do all this was Paul Butler's job, and it was so complicated that it took him six years to do it. "It's my Rembrandt," he says. "It's as close to great art as I'll ever get."

The code was so horrifyingly complex that it took four or five hours of computer time on the standard desktop computers of the early 1990s to analyze a single five-minute observation. This seemed somewhat excessive to other astronomers. At one point Marcy and Butler gave a talk at Harvard, describing the technique. They were almost literally laughed off the stage for the Rube Goldberg–like nature of their system. It may not be a coincidence that just about the entire Harvard astronomy department is on Butler's list of scientific scoundrels.

Geoff Marcy, though, has a certain sympathy for those who thought he was crazy. Nobody else had ever tried to make measurements this precise, so nobody had been forced to probe the characteristics of the spectrometer as if the instrument were part of the question being answered. He and Butler had tried at the beginning to get around such a daunting proposition, too. "Paul and I were dragged into this computation-intensive technique kicking and screaming," he says.

There remained one final insight that would eventually guarantee Marcy and Butler's status as the world's greatest planet finders. This was Geoff Marcy's realization that they had to look at a lot of stars to give themselves the best possible chance of seeing something. Maybe lots of stars in the galaxy had a solar system like ours, with a nice juicy Jupiter orbiting at about 450 million miles out and completing an orbit every eleven years. Maybe not. Bruce Campbell and his colleagues had a sample of twenty stars; it would later turn out that about 3 percent of the stars in the Milky Way have planets in the Jupiter range or somewhat bigger. Out of a sample of twenty stars you'd expect to see between zero and one such planets. The Canadians were unlucky and happened to see zero.

Nobody knew this in advance, of course, so Marcy and Butler put together what they considered a reasonably sized sample of 120 nearby stars—the 120 brightest stars that met their strict criteria for inclusion. The stars had to be similar to the Sun—a little hotter or a little cooler, perhaps, but about the same size. Their targets could be part of double-star systems but only if the pair was widely separated. Planets may not be able to form, and certainly won't stay in stable orbits, if they're constantly dodging an extra star.

Finally, their targets had to be stable. Many stars pulsate, beating to the rhythm of roiling gases deep inside. Their surfaces swell and subside, in and out, almost as though they were doing deep breathing exercises. But how do you tell the difference between a star that's moving toward and away from you and a star whose *surface* is moving toward and away from you? You don't. Marcy and Butler knew that their job was hard enough without adding an element of possible confusion. When you're doing very difficult observations, pushing your equipment to its absolute limits and trying to extract information from a complicated, confused mess of data, you're always in danger of thinking you see something when it isn't really there, as Schiaparelli did with the canals on Mars.

The one thing Marcy and Butler wanted to avoid at all costs was announcing they had found a planet, then realizing they had made a mistake and having to retract the discovery. Even worse would be announcing they had found a planet and then having someone else discover they'd made a mistake. The former would be deeply embarrassing; the latter would be humiliating. Both events had happened before in the planet-hunting game. As a result, said Butler, "the whole thing has a feel of snake oil." They were determined to avoid that feeling.

Chapter 4

OOPS, NEVER MIND

Andrew Lyne sits at his desk, utterly still. He stares at his computer screen, numb, barely able to think. The winter sun sets early in England, and afternoon has long since passed into evening, but Lyne is oblivious. The Jodrell Bank Radio Observatory is officially closed for the Christmas holiday. Lyne thought this would be a good time to get some work done in the office without interruption. He has been extremely busy for the past several months, and no wonder: Earlier in the year, midway through 1991, he had published a paper in the journal *Nature* announcing an astonishing discovery. Lyne and several colleagues had found a planet orbiting a star beyond the Sun.

Finding a planet was a big deal, of course. It was 30 years since Frank Drake had conceived the Drake equation, 60 years since astronomers had agreed that planets must abound in the Milky Way, 391 years since Giordano

Bruno was burned for insisting on the existence of other worlds. In all that time, nobody had found a planet anywhere beyond Pluto. Now Lyne had.

This planet was not exactly what Bruno or Drake or anyone else had had in mind. The star it was orbiting wasn't Sun-like. It wasn't even starlike. It was a neutron star, the burned-out husk of an ordinary star that had blown itself to bits. When a massive star runs out of fuel, it can no longer generate enough energy to remain inflated against the force of gravity. It collapses, then rebounds in a gigantic explosion so powerful that the star, now called a supernova, briefly outshines an entire galaxy of a hundred billion normal stars.

When the dust fades, all that's left is a swirl of hot gas—the Crab Nebula is one spectacular example, created in a supernova explosion observed by Chinese astronomers in 1054—and a neutron star. Neutron stars are the real-world answer to the textbook question of what would happen if you eliminated all the empty space in atoms. Neutron stars are not made of ordinary matter but of atomic nuclei packed right next to each other.

This isn't the sort of place you'd expect to find planets, since an explosion as bright as a galaxy would presumably vaporize everything for a vast distance around it. Nature isn't obliged to satisfy human expectations, however, and a planet orbiting a neutron star is what Lyne had found. This was a special class of neutron star known as a pulsar. For some reason pulsars emit tiny beeps of radio energy as they rotate, one beep per pulsar day (though the day may be a millisecond and a half long). One plausible theory argues that pulsars have powerful magnetic fields that funnel beams of radio energy out their magnetic poles and that these beams sweep across the Earth once every rotation, like the beams of a searchlight sweeping across a ship at sea.

Whatever the cause, these blips of radio energy are almost perfectly regular—so regular that when the first pulsar was picked up here at Jodrell Bank in 1967, Jocelyn Bell, the graduate student who found it, and her advisor, Anthony Hewish, who received the Nobel Prize as a result, thought for a while that they had solved the Drake equation. Here was a signal so perfect in its timing that it might plausibly be from an alien civilization.

It soon became clear that pulsars are a natural phenomenon, especially since astronomers quickly found several others in different directions. Unless the E.T.s had a galaxy-wide set of broadcast standards, this was strong evidence that pulsars, while bizarre and fascinating, were not signals from aliens. Within a few years astronomers who monitored pulsars realized that they weren't perfectly timed after all. Every so often one would begin sending out blips faster or slower than it had before. Astronomers speculated at first that the changes might be caused by planets. If you have a planet orbiting an ordinary star, pulling it back and forth, you'd expect, as Geoff Marcy and Paul Butler did, that the star's light would rise and fall in wavelength as the light was stretched and then compressed by the motion; this is what Marcy and Butler look for. In a pulsar you'd expect that the timing of pulses would change as the distance between pulsar and Earth varied. As the pulsar approaches the Earth, the radio energy has a shorter distance to go each time, and the pulses will come closer together; as it recedes, they should get farther apart.

As it turned out, the changes astronomers saw in the early seventies were irregular and unpredictable. No planet would make a pulsar behave this way. Eventually, theorists made a convincing argument that the changes were caused by starquakes shaking the neutron stars and accelerating their rotation rates.

That's not what Lyne had seen, however. His pulsar, known as PSR 1829-10 (the numbers are its celestial latitude and longitude), was sending out its blips of radio energy faster for a while, then slower for a while, then faster again, in a repeating pattern. It was just what you'd expect if there was really a planet out there. The actual observations weren't nearly this simple, though. To start with, the Earth isn't standing still; it's moving in its own orbit. If you don't allow for that, you'll observe a one-year variation in just about everything in the universe, caused by the Earth's motion alone.

Allowing for the Earth's orbit isn't so simple. To do so you have to know precisely where in the sky the pulsar is so that you know its angle with respect to Earth's orbit. Are we moving directly toward the pulsar or obliquely toward it, or what? Pulsars are generally too dim to be seen with optical telescopes, and the radio antenna Lyne and his colleagues were using—the giant 250-foot-diameter radio telescope at Jodrell Bank—couldn't tell them the pulsar's position to an accuracy of better than a tenth of a degree. The pulsar could be anywhere within a circle about half the size of the full Moon.

The way pulsar observers get around that problem is by taking an educated guess as to the pulsar's position, correcting for Earth's orbit, then watching it for a while. If the pulsar's timing varies in a regular way over a one-year period, its presumed position is adjusted and then recalculated, until the variation is eliminated.

Lyne couldn't eliminate the variation. It had to be a real motion in the pulsar, not just an error in fixing the position. The pulses were varying by between 8 and 9 milliseconds from the average interval of 330 milliseconds or so. Since radio waves move at the speed of light, something was pulling the pulsar 8 light-milliseconds closer than average to the Earth, then 8 light-milliseconds farther away, back and forth. Whatever was pulling on it had to be an object a few times Earth's mass in an orbit six months long.

Lyne and his colleagues were suspicious about the six months. When you see something that's an even multiple or an even fraction of Earth's orbit, you immediately suspect you might be picking up the Earth's motion in some way. But it wasn't *precisely* six months—it was two or three days off. They rechecked the calculations, and it all held up. They finally felt confident enough to send the paper off to *Nature*, which sent copies to several other astronomers for comment. These referees, as they're called, realized what a groundbreaking discovery this was and looked over the paper carefully. They finally agreed that it was sound. *Nature* published the report.

Lyne's paper, written in collaboration with Matthew Bailes and Setnam Shemar, caused a predictably and appropriately enormous stir. Maybe the planet wasn't where any planet was expected; maybe it couldn't possibly harbor life but was a planet. It counted. The ancient quest was over. Just a few months after Lyne's report appeared, Alex Wolszczan, now at Penn State, announced that he, too, had found planets orbiting an entirely different pulsar.

Wolszczan had been watching his pulsar for nearly a year. He'd seen evidence in the timing of pulses that there was some sort of orbiting companion, but he wanted to be absolutely sure of what he had. He asked a colleague, Dale Frail, who worked at the Very Large Array radio telescope in

New Mexico, to monitor the pulsar as well. He says, "It became clear that there was not one but two Earth-size objects orbiting the pulsar, one with a period of sixty-six days and the other with a period of ninety-eight."

At about this time, rumors began to fly that Andrew Lyne had discovered planets around a pulsar, too. "It was a tremendous surprise," says Wolszczan. "I considered rushing ahead to publish my data so I could be first, but I decided I wanted to be absolutely sure. I was willing to be second rather than take a chance on being wrong."

He finally went public a few months after Lyne did. With two examples of planet-bearing pulsars, the astronomical community had to take the phenomenon seriously. The president of the American Astronomical Society, John Bahcall, an astrophysicist at the Institute for Advanced Study in Princeton, New Jersey, decided that this unexpected set of findings merited a special session at the society's next winter meeting, which would be held in Atlanta, Georgia, in January. Both Lyne and Wolszczan were invited to present their results to the largest regular gathering of astronomers on the planet.

Now it was late December; Lyne was in his office, catching up on a backlog of work and getting organized for the big conference less than two weeks away. Suddenly, as he sat there thinking about the planets and what he would say about them, Lyne was struck with an appalling thought. When he and his coworkers had first tried to nail down the pulsar's position, they had properly included a correction factor for the Earth's orbit. The Earth's orbit is elliptical—very nearly, but not quite, a perfect circle. On the first go-around, though, which is always approximate, it's normal to pretend the Earth moves in a true circle. It makes the calculations a lot easier. Later on, you substitute the actual elliptical orbit.

"I suddenly realized," he recalls, "that we had never gone back and done this recalculation." He did it now, dreading what might happen. "The planet," he says, "disappeared." It had never been there in the first place. It was all a mistake. "The next few days," he remembers, the distress evident in his voice years after the event, "were a pretty awful time. We'd received a great deal of publicity. A lot of people had listened to what we'd said with great faith and trust. You couldn't blame the referees, because all they had to go on was the information in our paper. I had to break the news to my collaborators and then to the observatory.

"And then I had to ring up John Bahcall, who had offered us this wonderful public forum. 'I have this problem,' I remember telling him. He offered me the chance to go anyway and explain what had happened. This was not what I wanted to do. But I had to let the world know somehow, and the sooner the better. John encouraged me to go ahead."

I was in Atlanta to cover the meeting for *Time* magazine, and I knew John Bahcall from having interviewed him for other stories. The morning of Lyne's and Wolszczan's talks, Bahcall came to the pressroom and saw me leafing through some notes. "I want to encourage you in the strongest possible terms to come to the pulsar planets session," he said gravely. "You will see a shining example of what science is all about."

When I showed up at the hotel ballroom reserved for the session, every seat was filled, and the back of the room was crowded with standees. More than a thousand astronomers—nearly everyone who had come to the conference—were there. Rumors had been flying back and forth by E-mail for the past two weeks, and by word of mouth through the conference for the past couple of days. Everyone had a good idea of what was coming. Every one of the astronomers present had had nightmares of being in Lyne's shoes. Every one of them wondered how Lyne would handle it. They didn't need John Bahcall to tell them that this would be a memorable event.

Andrew Lyne stepped up to the microphone and said in a quiet but firm voice, "I originally planned to discuss an important discovery here. Instead, I must issue a retraction and make an apology." He explained carefully and quietly exactly what had happened. When he was finished, every astronomer in the room stood up, and the room erupted in deafening applause. It lasted more than a minute.

"I have to admit I was very surprised," Lyne told me. "I was admitting a mistake that I never should have made. People were very kind, which certainly made things a lot easier. But I really didn't think the applause was justified." John Bahcall cornered me right after the talk was over, as determined as a presidential press secretary to put the right spin on Lyne's performance. "I want you to know that Andrew Lyne's talk was the most honorable thing

I've ever seen. A good scientist is ruthlessly honest with him or herself, and that's what you've just witnessed."

John Bahcall is an eminent and respected astronomer. Paul Butler has referred to him as one of the "gods of astronomy." Nevertheless, neither Butler nor Geoff Marcy wanted to receive the sort of praise Bahcall had bestowed on Andrew Lyne. In order to make an honorable admission of error, you first have to make the error publicly. Scientists—not just astronomers but biologists, geologists, chemists, and others—make errors all the time. Properly practiced, though, the process of science includes constant self-criticism and second-guessing. Could what I think I'm seeing be caused by something else entirely? Is my statistical analysis rigorous enough? Have I screwed up here? Wolszczan had been more careful in his analysis, and his discovery is now accepted.

Lyne hadn't found his mistake in time to keep it quiet, but at least he'd found it before someone else did. What Butler and Marcy feared even more was that they'd someday be mentioned in the same sentence with Peter van de Kamp.

Van de Kamp, too, had found a planet, one that was orbiting an ordinary nearby star. It happened at about the same time that Frank Drake was creating his equation, as the Swarthmore College astronomer was nearing the end of a long career. Van de Kamp had been searching for planets around nearby stars since 1937, using the twenty-four-inch refracting telescope at Swarthmore's Sproul Observatory. Where Walker, Campbell, Marcy, and Butler would one day look for the wobbling of a star's spectrum, van de Kamp looked for its physical motion, back and forth across the sky, rather than toward and away from the telescope.

It took him a quarter of a century, but he finally got his planet. The object was about 1.6 times as massive as Jupiter, moving in an orbit comparable to Jupiter's, around Barnard's Star, which lies six light-years from Earth. Like Lyne, van de Kamp was invited to present his results to the American Astronomical Society. The new planet was accepted by other astronomers. A few years later, van de Kamp announced new observations that confirmed the discovery. The planet—maybe two planets, van de Kamp now suggested—became part of the conventional astronomical wisdom. It appeared in textbooks, and still does, more than three decades later. High school and college students taking introductory astronomy courses still learn about the planets around Barnard's Star.

Except that there aren't any planets around Barnard's Star. Van de Kamp was simply mistaken. The man who proved it is George Gatewood, now the director of the University of Pittsburgh's Allegheny Observatory, but then — in the early 1970s — a graduate student at the same institution.

The problem with van de Kamp's measurements, most astronomers now agree, was one that is inherent in refracting telescopes. As previously explained, a refractor uses lenses to focus and gather light, whereas a reflector uses a curved mirror. In the former, light has to pass through a thickness of glass. Unless light enters the glass absolutely head-on, i.e., at a 90-degree angle to the surface, it's bent slightly as it enters and bent again as it leaves. And light of different colors is bent at slightly different angles — red is bent more and blue less, and green somewhere in the middle. If the light is made up of many colors, as sunlight and starlight are, the glass will bend each of its colors by a different angle; a beam of white light comes out as a rainbow. (There is no such problem with reflecting telescopes because starlight simply bounces off the surface of a curved mirror. It never passes through a glass lens, so all its colors arrive at the same place at the same time.)

In a telescope lens, this means that blue and red light focus at slightly different distances. That tends to blur an object's shape. When you're trying to measure the precise location of a star on a photographic plate, this can be a big problem — and evidently it was van de Kamp's problem. The telescope at Sproul had had its lens remounted during his observing program; the focal length was changed just a fraction, and that change showed up as a minute shift in position for the stars. "It was kind of suspicious all along," says Gatewood. "The planets Peter discovered all had periods of about twenty-five years, and the lens was changed twenty-five years after he started observing."

Van de Kamp refused for many years to accept the fact that he was wrong. A comment he made to Gatewood shortly before his death suggests that he eventually capitulated. "George," Gatewood remembers him saying, "you must not spend your whole life looking for other people's systematic errors. You must make some of your own."

In effect he was saying you can't make discoveries if you're not putting yourself on the line and taking risks. Any thoughtful scientist would agree and would applaud what appears to be van de Kamp's implicit statement that he himself had made errors. It's better to catch them yourself, though, as Lyne did, and better still to catch them before publication. A planet-hunter himself, Gatewood says, "It's happened more than once that I've thought I had something." When that happened, he generally sent the data off to David Black, head of the Lunar and Planetary Institute in Houston, Texas, an old friend and a tough critic of flimsy claims. So far, Black had always responded in a way that made Gatewood think again and eventually back off: "George, are you *sure*? You don't want to become known as the Peter van de Kamp of your generation."

Chapter 5

OTHER WORLDS

For most of the eight and a half years they worked on their project, Geoff Marcy and Paul Butler didn't have to worry about being compared with Peter van de Kamp. They couldn't be wrong because they hadn't found anything. They hadn't even tried. They'd been making observations right from the beginning, but the effects they were looking for were so subtle that no human eye could detect them. Only a computer could tease out the signal of a planet from their complex, messy data. But there was no point in analyzing the data at first because at best their technique was sensitive enough only to find large, Jupiter-size planets. Jupiter takes eleven years to go around the Sun, so you wouldn't expect to be confident you'd found a Jupiter in much less time than that. Ideally, you want to see a good fraction of a full orbital cycle before you start talking about planets.

But even after six years into the project, with their software finally in place, there wasn't much point in analyzing their data because it would take far too much computer time. The astronomers who laughed them off a Harvard stage may not have been polite, but they weren't stupid. So Marcy and Butler kept toiling in obscurity, storing megabyte upon megabyte of star images on magnetic tape, hoping that someday they'd be able to do something with them.

Worse yet, their deblurring software had revealed problems with the Hamilton Spectrograph itself. It might be one of the world's finest, but it wasn't fine enough. It didn't work as well at the observatory as it did on paper. At best, the astronomers could measure a star's motion to within twelve meters per second—a hair better than anyone in history but nowhere near the precision they needed. They couldn't find something as small as Jupiter with their system; the motion a Jupiter imposes on its star is twelve meters per second, making their errors as big as their measurements.

Marcy and Butler made all of this known to Steve Vogt, who finally freed up enough time to go up on the mountain and deal with it. Vogt worked the way an electrician would to find the source of a short circuit: He traced the path a beam of light would take from the moment it entered the inner sanctum, looking at each point of reflection and each lens for a source of blurring.

He already suspected, though, where the problem was. Since it is essentially impossible to make all the elements in an optical system perfect, the Hamilton has a lens designed to correct any blurs or aberrations that have been introduced by imperfections in the rest of the optics. The device is called a corrector. Vogt took it out of the spectrograph and discovered that it suffered from "slope errors," a kind of optical imperfection. "It was," he says, "an incorrect corrector."

Once the corrector was rebuilt and the entire spectrograph refurbished, in November 1994, Butler and Marcy had to rewrite their code to take into account the physical improvements. Finally, in April 1995, they were sure they were ready. They took a reading on the one star in the universe whose motions, pulsations, and undulations were known in intimate detail by astronomers. "We were able to recover the Sun," said Butler, "stable at levels of just three meters per second. That's when we knew the code worked." Now all they needed was a computer fast enough to process their messy software at a reasonable rate of speed.

Marcy and Butler had known since the Lyne and Wolszczan episode that they wouldn't be the first to discover extra-solar planets. At least one pulsar in the Milky Way had at least two planets orbiting it. (Since Wolszczan's initial discovery, astronomers have found two other pulsars that might have planets and another—in an astonishing demonstration of how precisely such things can be measured—which has something orbiting it that is about the size of a large asteroid.)

Still, planets around normal stars were the real quarry, the only thing that really counted in solving the Drake equation. It is conceivable that some bizarre sort of life can exist on pulsar planets or that an advanced civilization could somehow survive a supernova explosion. Nobody thought it likely enough to worry about, though. Wolszczan's planets were the first, but because they probably had no bearing on the question of life, they went into the record book with an asterisk beside them, signifying that in a sense they didn't count, just as Roger Maris's sixty-one home runs in 1961 didn't count: He beat Babe Ruth's record of sixty in a season, but Maris had the advantage of a longer season.

Another possible planet discovery had been reported back in 1989, but this, too, didn't really count. A Harvard astronomer, David Latham, found something circling a star known as HD114762; the object was certainly dark and a lot less massive than a star, but neither Latham nor anyone else could be sure of its precise mass. The thing was at least nine times as massive as Jupiter and possibly much more. Latham decided, conservatively, to label the thing a probable brown dwarf—a star that isn't quite big enough to shine—rather than a planet.

Toward the end of September 1995, though, Geoff Marcy began to receive disconcerting E-mail messages. Someone else had made an important discovery that would not make him happy. A major announcement was coming any day. "Right at the end of September, Geoff heard that it was about to go down," says Butler, the son of a cop from *Dragnet* country, in perfect Joe Friday deadpan. "It went down," says Butler, "on October 6." And it came from a direction nobody would have expected.

* ○ *

Michel Mayor never expected to be the first scientist to find a planet orbiting a Sun-like star. The University of Geneva astronomer wasn't even
looking for planets, since he knew he didn't have a prayer of detecting them.
Like Marcy and Butler, he was looking for the telltale forward-and-back motion that would betray the presence of an unseen object going around a star.
But while the California astronomers could now measure these motions to
an astonishing accuracy of plus or minus three meters per second, Mayor's
equipment was much less precise. His errors were around thirteen meters
per second, or about what Marcy's and Butler's had been before Steve Vogt
rebuilt their spectrometer. You could never find a Jupiter with that kind of
precision.

But Mayor was not looking for planets but for low-mass objects, a much
broader category that takes in everything smaller than a star, and much easier to find—though hardly easy. David Latham's brown dwarf is a perfect example. Theoretical astrophysicists had believed for years that there was
something intermediate in size between Jupiter, which is about one-tenth of
1 percent the mass of the Sun, and the smallest stars, which are about 8 percent of a solar mass—about eighty times bigger than Jupiter. Why, they reasoned, would Nature have left such a huge gap in sizes? The gap should
logically be filled with something bigger than planets but smaller than stars.
Since the smallest, dimmest stars are known as red dwarfs, these missing
bodies were dubbed brown dwarfs by a NASA astronomer named Jill Tarter;
they put out some heat but not even enough to glow a dull red. The search
for brown dwarfs had been proceeding in parallel with the search for planets,
and here, too, there had been several false detections before David Latham's
successful one.

I met Mayor at the International Astronomical Union's major conference
on the topic of bioastronomy—the Drake equation. The conference is held
every five years, and this time the site was a conference center on the isle of
Capri. Given the setting, everyone with even a remote connection to the
topic of life on other worlds made an extra effort to be there.

Frank Drake showed up, and so did Philip Morrison. Geoff Marcy and
Paul Butler came, with wife and girlfriend, respectively. (Marcy's wife, Su-

san, a chemist, attended most of the scientific sessions; Butler's girlfriend, Nicole, an investment banker, mostly went shopping at Hermés, Gucci, and the other frighteningly expensive shops lining the broad sidewalk that serves as the main street of the town of Capri.) Seth Shostak, the astronomer who paced his bedroom floor waiting to hear if alien signals had really been detected in Australia, was there. Carl Sagan was expected, but he was already, six months before his death, too ill to travel. A handful of Nobel Prize winners showed up as well.

Mayor finally found some time to talk over lunch during the third day of the conference. He wears a thick, full, reddish beard, which balances a receding hairline. (Evidently, and for reasons that remain mysterious, this is the standard look for people who find planets.) We sat in a sunny courtyard surrounded by lush purple bougainvillea and ate Caprese pasta while he told me how he ended up justifying Giordano Bruno's martyrdom.

Mayor had been thinking about low-mass objects for a long time, he said, but became actively interested in 1989 when David Latham discovered HD114762. Both Latham and Mayor had noticed that the star's spectrum varied in a period of eighty-four days, and mindful of earlier false detections, they checked with each other to make sure they weren't making some foolish mistake. Having made sure, Latham went ahead and published, which is why he is credited with the discovery.

This episode convinced Mayor that finding brown dwarfs was something he could do. He called in an instrument maker named André Baran—a sort of French Steve Vogt—to upgrade his own spectrometer. Rather than fool around with iodine cells and horribly complex computer programs, Baran and Mayor decided that they would make the spectrometer itself as stable and free from distortions and errors as they possibly could. They fitted their spectrometer with two optical fibers, the same kind that carry phone conversations and Internet data across continents. One fiber was connected to the telescope at the Haute-Provence observatory in the Pyrenees mountains, on the French-Spanish border. The other was connected to a reference lamp; that way, Mayor could be sure that starlight and reference light were entering the spectrometer exactly the same way each time. "This," he said, "is how we have been able to achieve an accuracy of thirteen meters per second."

No one knew how common brown dwarfs would actually turn out to be in the universe, no matter what the theorists believed. (It's a maxim in astronomy that an observer who is wrong as much as 10 percent of the time is hopeless, but a theorist who is wrong as *little* as 10 percent of the time isn't using enough imagination.)

Like Geoff Marcy, Mayor decided to maximize his chances of a discovery by looking at a large sample of more than one hundred stars. Among the objects in his sample was a nondescript yellowish star called 51 Pegasi, in the northern constellation Pegasus, 51 Peg for short. Marcy and Butler hadn't put the star on their list because the star atlas they used, a reference known as the Gliese catalog, described 51 Peg as a sub-giant—a star sufficiently un-Sun-like that it didn't make the cut.

But Mayor didn't really care. He wanted to find low-mass companions no matter what kinds of stars they were orbiting. He wasn't that interested in the Drake equation. So 51 Peg went on the list, and in 1994, six years after Marcy and Butler began their project, Mayor began observing.

Working with graduate student Didier Queloz, Mayor began ticking off the stars on his list, visiting each one about every two months. In the fall of 1994 they noticed that one or two measurements were discrepant. When you have a new instrument, you expect problems, so you go back and check. While Mayor was at a conference in Hawaii, his colleague Queloz checked. The other stars looked fine, but 51 Peg still seemed to show a discrepancy. Queloz sent Mayor an E-mail containing the new observations and a question: "What do you think?"

What he thought was that the numbers were crazy. If there was something orbiting 51 Peg, it was like nothing anyone had ever seen or even imagined. The object had a low mass, all right. It was about half the mass of Jupiter, a bit bigger than Saturn. But this thing was racing around 51 Pegasi so fast that its year was only four days long. That's how long it took for the star's wobble to complete one full cycle. Once they knew the object's orbital period, Mayor could calculate its distance from its star. It was closer to 51 Peg than Mercury is to the Sun—*seven times* closer. You could imagine a tiny chunk of rock in such an orbit or maybe a burned-out comet or even an errant moon that had somehow been torn loose from its planet. But a planet bigger than Saturn?

Yet the object would have to be as big as a super-Saturn to explain why Mayor and Queloz could detect it with their relatively insensitive system. In our solar system, Jupiter makes the Sun move back and forth at thirteen meters per second, but that motion is based not just on the planet's mass but on its location, 450 million miles out. Closer in, Jupiter would have more leverage. A hundred times closer—for this whatever-it-was was whipping along just four million miles above the fiery surface of 51 Peg—and you give it a whole lot of leverage. Even at half of Jupiter's mass, the object was yanking its star around at about thirty meters per second.

This was all so absurd that at first Mayor was sure they were being fooled by some other effect masquerading as a planet—some sort of pulsation in the star's atmosphere or some strange effect related to the star's rotation. Yet it didn't really look like either to his practiced eye. What it looked like was a planet. Obviously, they needed to take a closer look, and they had to be circumspect about it. So when they applied for additional telescope time, they didn't specify what they thought they'd found. They just said, "We have a candidate that merits further observation." They had to wait until 51 Peg reappeared over the horizon, but finally, in the first week of July 1995, they looked again. There was the variation, just as it had been before.

As Sherlock Holmes once observed, once you've ruled out the impossible, whatever is left must be true, no matter how improbable it seems. A large planet in such an orbit was wildly improbable, yet it seemed to be true. Mayor and Queloz weren't taking any chances, though. They began checking their work intensively, going over the calculations, reviewing their assumptions, trying to think of any way they could possibly be fooling themselves. "It's a difficult thing to decide you've done all you can, that you're ready to leave your office and go public," said Mayor. By the end of August they'd concluded that it would have to be now or never. They wrote up their results and submitted a paper to *Nature*.

Most major scientific journals insist that any results announced in their pages must appear there first. If they show up somewhere else—a newspaper, for example, or a magazine like *Time* or *Newsweek*—the journal will unceremoniously yank the article from the publication schedule. Because scientific journals have much higher prestige than consumer publications, scientists take pains not to talk to the press about their work until after it appears officially.

They will, however, talk to their colleagues. Mayor had been invited to participate in a scientific workshop on low-mass stars to be held in Florence, Italy, in early October. There would be three hundred people at the meeting, experts in many sub-specialties of astronomy. "I said to myself, okay, this is a good place to talk about our results," he recalled. If the assembled astronomers couldn't find any flaws in his and Queloz's reasoning, they were probably on firm ground.

As the conference date approached, Mayor still hadn't heard from *Nature* as to when or even whether his paper on 51 Peg would appear. He wrote to them and asked if they had any objection to the October presentation. Finally, the night before he was to speak, Mayor received a reply. One of the referees wanted to reject the paper on the grounds that the observations simply weren't convincing. The other two, however, had only minor questions—good ones, Mayor felt. They wanted Mayor and Queloz to rewrite the paper in order to address these questions and persuade the third referee (they are usually anonymous) to vote yes.

Nature also responded to the more pressing question of whether Mayor could talk about his work. Yes, he could do this, but he couldn't present it directly to the press. He couldn't grant interviews. *Nature*'s editors had some leverage over Mayor, but they weren't naive enough to think that any reporters present at the meeting would censor themselves. So they gave Mayor a printed statement to hand out. It said, in effect, "Don't report on this, but if you do, be sure to include the proper caveats." The warning had about the same effect as the line, "Pay no attention to the little man behind the curtain!" in *The Wizard of Oz*. If the press was being told not to write about it, it must be important.

So Mayor presented his data and then his interpretation. He offered three possibilities: The apparent wobbles could be due to stellar oscillations, a rotational quirk, or a low-mass object. "Some members of the audience thought that the third was the least probable idea," said Mayor, "but we had strong support from others. When we finally left the room, I had a strong sense that nobody had presented any very strong arguments against our conclusions."

Word got out immediately, of course. Many of the astronomers who had been in the audience E-mailed their colleagues—Geoff Marcy among them—and reporters didn't give *Nature*'s warning even a moment's consideration. They grabbed every astronomer they could get their hands on in Florence and then, as the news spread, at universities and observatories around the world. Mayor wouldn't talk, however. He abided by *Nature*'s injunction. "It was an uncomfortable period," he said. "Many astronomers were discussing these things with the press, and we were not. Many of them didn't even have any information. They got the name of the star from the newspaper." The paper that formally announced Mayor and Queloz's discovery, freeing him to talk, would finally appear on November 23, a month and a half after the Florence conference and long after the initial interest had died.

Back in Berkeley, meanwhile, Geoff Marcy and Paul Butler were wondering what was going on. Marcy had immediately sent Mayor a congratulatory E-mail, but now he and Butler were suddenly getting E-mails themselves, and faxes and phone calls, from reporters who knew that they were hunting for planets. If this guy Mayor wouldn't talk, maybe Marcy and Butler would.

Once they'd heard the details, though, the American astronomers were convinced that Mayor must be wrong. No planet could be this strange. They also knew they were better equipped than anyone to prove it. Thirty meters a second? Piece of cake, even if it would take a lot of computer time to analyze.

Purely by good fortune, Marcy and Butler happened to have four observing nights reserved at the 120-inch telescope right at that time—just enough to see 51 Peg through one cycle of whatever was going on. Over that period they made twenty-seven observations of the star.

They didn't have much time for sleep, even when 51 Peg was below the horizon. Data analysis was a huge problem to begin with, and now, because of all the interest they were getting from the press, Marcy and Butler were trying to analyze the data as soon as they came in. Butler would grab the observations as they came off the telescope, like an old-time newspaperman ripping copy off the teletype. He'd E-mail them to Eric Williams, a graduate student who was helping out; Williams would do some preliminary analysis, then shoot the streamlined data back to Butler. "We had three different computers going at once," Butler recalls.

By the evening of the third night they knew the star really was varying. By the fourth morning they could tell that the light curve was snaking up and down just the way Kepler said it should. Just before they drove off the mountain, the truth dawned on Butler. "I thought, 'Jesus, maybe Mayor's right.' "

On the way back to Berkeley, he and Geoff Marcy agreed that they'd get some sleep, then meet later to see where they stood. They also discussed how they would handle things if Mayor really did turn out to be right. What they decided was that if they proved Mayor wrong, there was no point in making a public fuss. They'd communicate their findings through proper scientific channels. But if they confirmed his discovery, they'd make an announcement.

Douglas Lin, an astronomical theorist at the University of California, Santa Cruz, was perhaps the first person besides Marcy and Butler, their collaborators, and their families who found out which it would be. He had heard about Mayor's discovery from a friend who'd been in Florence, and about how skeptical everyone was about the purported planet's ridiculous orbit. Lin knew that his friend Geoff Marcy would probably have some information. As it happened, Marcy had just come down from the mountain. "He was *very* excited," Lin recalls. "I asked, 'Is Mayor correct?' Geoff said, 'Absolutely. I've just confirmed it. I can't stay on—we're doing a press conference in half an hour.' "

Marcy and Butler weren't quite prepared for what happened after the press conference. "When I got to my office," Marcy says, "the phone started ringing, and it didn't stop for weeks. We were inundated from then on." Ted Koppel grabbed them for *Nightline*. Local TV stations fought over who would get them first. Radio and newspaper reporters from around the world wanted—needed—interviews right away.

The excitement of instant celebrity paled for Butler, though, before the simple fact of the discovery. This was a man who had grown up worshiping Giordano Bruno rather than Mickey Mantle, the man who would a few months later, en route to the July 1996 bioastronomy conference on Capri, make a pilgrimage (dragging Nicole along with him) to the site of Bruno's immolation in Rome. "For the next two weeks I couldn't sleep," he says. "The universe was screaming in my head, 'There are planets out there!' "

Another voice was screaming, too: "Mayor got there first. His spectrograph is less sensitive than ours, and he's been doing this for only a year and a half, and we blew it." The Californians could easily have detected the planet around 51 Pegasi themselves, even before the spectrograph was rebuilt in 1994. But they would have had to be incredibly lucky, because no planet-hunter in his or her right mind would have deliberately looked for such a crazy planet. To find a Jupiter, you take a look once every few months, and over several years a pattern appears—the star is moving toward you, toward you, toward you . . . now it's slowing down, now reversing, now it's moving away from you, away from you. . . . It wouldn't matter a bit if your observations were evenly timed because each leg of the planet's journey takes so long. With a 51 Peg, though, you could easily miss the planet entirely. You'd see the star coming, then going, then somewhere in between— a seemingly random jumble of motions unless you took enough observations to fill in the gaps, as Mayor happened to do.

So because Marcy and Butler were sensible and didn't bother looking for what couldn't be there, they missed finding it. But Marcy and Butler, at least (perhaps by necessity), made peace with their bad luck. "People ask me if I feel bad that we didn't find 51 Peg," says Marcy. "I really, really don't. In retrospect, we could have found it, or planets like it, if we'd analyzed our data instead of sitting on it. But I still don't think we made the wrong decision." If they'd known it would be so easy to find the first planet, they might not have worked so hard on the software and the spectrograph, to wring the last drop of sensitivity out of their system. They missed being the first to find a planet; instead, they created a machine that can churn out high-precision velocity measurements one after the other.

As he and Marcy fielded nonstop interview requests in the aftermath of the 51 Pegasi discovery, it wasn't just heroic thoughts of Giordano Bruno that raced through Paul Butler's head. Mayor had suggested during his presentation in Florence that there might be a second planet orbiting the star, one with a longer and more sensible orbit. "51 Peg could be the first example of an extrasolar planetary *system* [italics mine] associated with a solar-

type star," he and Queloz wrote in their formal report. The second planet
wasn't nearly as certain as the first, but Marcy and Butler were determined
that they'd be the ones to prove its existence if it was there. So the first thing
they did when they got back to the mountain was start monitoring 51 Peg as
closely as they could. "We stole a lot of guerrilla time on the telescope," said
Butler. He was up there for a month, almost nonstop.

They also had a second, more significant concern. The Berkeley astronomers had created a velocity measurement machine of unrivaled accuracy, but it was sitting idle. Mayor's discovery had made it clear that there
could be planets lurking, inscribed in magnetic scribblings laid down on
data tapes in writing only a computer could decipher and understand. At the
time, Marcy and Butler thought Mayor was working from the same list of
stars they were, so they envisioned a scenario where Mayor would find everything—all of those luscious planetary plums just hanging from the tree. He
could do it quickly because he didn't need to process the data very much at
all. Butler was damned if Mayor was going to get all those plums, but with
his slow computer he knew it would take three years to reduce his data. "I
was insane to know what was on those tapes," he says. "It was a bad scene."

He simply had to have more processing power. So Butler began cruising
around the World Wide Web looking for people to borrow from. He became
an astronomical Oliver Twist, imploring people to feed his hunger for processing power. Finally, two groups in the astronomy department lent him
their SPARC 20 workstations, and Sun Microsystems eventually donated an
Ultra, one of its fastest desktop machines. He essentially stayed in his office
crunching data, or on the telescope, around the clock for the next three
months. Nicole was not happy. They'd been planning to go to China, and
Butler told her "no way." She began bargaining—could they maybe go at
Christmastime? "I said, 'Forget it.' "

The first planet showed up about a month and a half later, in early December 1995. Butler had crunched his way through nearly half of the 120
stars on the list. One of them was a Sun-like star known as 47 Ursae Majoris
(47 UMa), in the Big Dipper. Butler's software refined the star's blurry spectrum as recorded by the Hamilton spectrograph, turning it into a sharp set of
lines. The software compared that spectrum with the spectrum of iodine,

imprinted on the starlight as it entered the spectrograph, then did it again and again for a series of observations stretching back several years. It determined that the star was moving toward and away from Earth, and plotted each velocity measurement as a dot on a chart. And in its final step it traced an undulating line through the dots—the telltale signature of a planet waltzing through space with the star. Butler had found his first planet.

"I showed it to Geoff," said Butler. "But Geoff is a very skeptical scientist, which is good. He saw the signal and said, 'So what?' " Marcy's objection was that the signal was only four times larger than the errors. It was a four-sigma detection. That meant the result was believable but not rock solid. Marcy wanted to be absolutely sure, so over the next few weeks Butler reanalyzed the data, adding in the newest observations of 47 UMa and putting them together in what he describes as "a morally, ethically blind way."

It was right in the middle of this reanalysis that Butler went to the office early one Saturday morning to check on the latest batch of calculations. He and Geoff Marcy had been there late the night before, feeding data into the computer, and he wanted to get the results and move on to the next star. "I saw that the processing run for a star called 70 Virginis was done. By now I was getting pretty good at reading a signal. This one was ugly." But he patiently ran his period-finding software, which found that the signal repeated every 117 days, and then the orbit-fitting code, which traced a curved line through the peppery dots.

This was the morning of the forty-sigma detection, the measurement forty times greater in accuracy than the margin of error. This is what set him into an altered state of consciousness—the state in which he phoned Geoff Marcy, babbling, and dragged Marcy and his wife away from their New Year's Eve shopping trip and into the office, to say, "Oh, my God."

* ○ *

The American Astronomical Society has two major meetings every year, one in January and one in June. By June, most universities are not in session and astronomers are frequently off observing or traveling, so the June conference is often sparsely attended. The winter meeting, by contrast, is always full. It may be pure coincidence, but it may also be due to the fact of a bigger audience of both scientists and reporters that major findings seem to be presented more often in January than in June. Andrew Lyne and Alex Wolszczan gave their talks on planets around pulsars in January 1992. Margaret Geller of Harvard, in January 1986, presented the shocking news that galaxies are not spread uniformly through the universe but are gathered into huge, thin structures as though glued to the surfaces of giant soap bubbles. John Mather of the Goddard Space Flight Center announced that the Cosmic Background Explorer Satellite had confirmed a major part of the big bang theory in January 1991.

The 1996 winter meeting of the American Astronomical Society took place only two weeks after Paul Butler and Geoff Marcy found themselves staring in shock at 70 Virginis's textbook-perfect light curve on Butler's computer screen. As happens at most scientific conferences these days, the AAS press officer, Steve Maran—a professional astronomer himself who works at Goddard Space Flight Center—had come up with a series of press conferences to help reporters find what might be interesting out of the thousands of papers being delivered over the four-day meeting.

The day the conference began, Maran told me, "You'll see on the program that we have a press conference scheduled for Wednesday with Geoff Marcy of San Francisco State. I strongly urge you to be there." What was it about? "I can't tell you." Steve Maran is utterly discreet and doesn't reveal scientific results, even in private, before they're officially announced. On the other hand, he isn't cruel. "What I can do is remind you that Marcy is the one who confirmed the planet around 51 Pegasi last fall, and he's been doing his own search for planets. You really should not miss this press conference."

I took the suggestion. Most of the press conferences that week were being held in a small conference room in the San Antonio Hilton hotel, the meeting headquarters. It was big enough to hold all the reporters at the conference, but by the time Wednesday arrived, word had spread that Marcy had a big announcement. Plenty of astronomers didn't want to wait for Marcy's formal talk the next day. They besieged Maran with requests to go to the press conference, and at the last minute he switched it to a medium-sized ballroom.

The place was packed. TV lights blazed from every corner. Astronomers and reporters sat patiently, then began shifting in their chairs as they listened to several preliminary reports about 51 Pegasi—how it might have formed, what it might be made of. Nobody wanted to hear it. This was old news. Finally, Geoff Marcy stepped up to the lectern. Paul Butler sat in the front row, beaming.

"After the discovery of 51 Pegasi," he began, "everyone wondered if it was a one-in-a-million observation." He paused for emphasis and surveyed the crowd. "The answer is . . . no. Planets aren't rare after all." He went on to show the light curve of 70 Virginis, printed out and projected on a giant screen at the front of the auditorium. The scientists in the audience immediately understood its significance. Then he showed the curve of 47 Ursae Majoris—for his initial "So what?" had disappeared with Butler's extra several weeks of reanalysis. Now Marcy, too, was convinced that there was a planet orbiting 47 UMa. Four months earlier, not a single extra-solar planet orbiting a Sun-like star had been known to science. Now, evidently, there were three.

Marcy wasn't finished. Having announced the existence of two new planets orbiting Sun-like stars, he then described their orbits. These planets weren't scalding oddities like 51 Peg B. (Following astronomical convention, a star's companion has the star's name plus the letter B, a second companion would be labeled C, and so on.) They circled their stars at respectable distances—equivalent to somewhere between Mars and Jupiter for 47 UMa B, between Venus and Mercury for 70 Virginis B. In the case of 70 Vir B, that implied a surface temperature of about 185 degrees Fahrenheit, which Marcy carefully pointed out was in the range where water could exist in liquid form. "And liquid water," he added for the benefit of the reporters present, "is a necessary requirement for life as we know it."

This was a perfectly accurate and respectable set of scientific statements. Marcy didn't say there *was* liquid water on 70 Vir B, only that there could be. He was careful to note that the planet, which is about six times as massive as Jupiter, is most likely made up mostly of hydrogen gas, as Jupiter is, and thus doesn't even have a solid surface (47 UMa B, about three times as massive as Jupiter, is probably also a gas giant). If life exists there, it would have to be permanently airborne—bizarre balloonlike organisms, perhaps, for which there are no analogues on Earth. Life might also conceivably exist on a giant moon circling the planet, but there's no evidence at all for such a moon.

The real significance of 47 UMa and 70 Vir, Marcy took pains to explain, was the implication that there were more planets out there to be found. Although Marcy didn't mention it by name, the Drake equation was implicit in what he was saying. With the discovery of a planet around 51 Pegasi, the term F_p in the equation, describing the fraction of stars that form planets, had been doubled. Science had only known of one star in the universe where planets existed; now they knew of twice as many. With Marcy and Butler's discovery, the number had been doubled again, from two to four. "Planets aren't rare after all," he'd said. If that was really the case, then astronomers should be finding more—entire solar systems, eventually, and even Earth-like planets someday.

The scientists in the audience understood this subtext, and when Marcy was finished, the moderator of the press conference, an astronomer with the Space Telescope Science Institute, made the same point in a more explicit and sweeping way. "What we are seeing," said Robert Brown, "is the culmination of intellectual history that began with Copernicus five hundred years ago." Sitting in the audience, Paul Butler smiled even more broadly. He had now been publicly compared with one of his astronomical heroes. "This isn't just a scientific event," said Brown. "We have entered a new period of human thought."

The press seized on quite a different point. Intellectual history was all very nice, and so were future discoveries. But Marcy had used the word "life." The reporters present were all careful to include Marcy's caveats in their stories, but they were also sure to feature the L-word very prominently. What could be more exciting than even the barest possibility of life on other worlds?

The Associated Press had a story on the wire within minutes. Soon afterward, phones in the pressroom began ringing with requests for Marcy and

Butler to give network TV interviews, right away if possible. One call was for me: The science editor of *Time* wanted to know if I thought there was a cover story in this. I looked around at the other reporters frantically trying to get hold of a phone to call in their requests for space on the front page. Yes, I thought there probably was. I checked with Marcy and Butler, who were out in the hall surrounded by a crowd of my colleagues, to make sure they would make time to talk to me. The story came out a week and a half later. The line on the cover said nothing about intellectual history. It read: "Is Anybody Out There?"

The phone kept ringing through the rest of the conference, and when the astronomers got back to Berkeley, it continued ringing. First the daily papers called, then the weekly magazines, then the monthlies. Network news crews got what they wanted in the first day or two, but then there was the BBC and the Discovery Channel and *Nature* and *Nova* and all sorts of other documentary crews. After a couple of weeks, Marcy and Butler let their answering machine take the calls. A month later, sitting at the controls of the 120-inch telescope at Lick Observatory in the mountains above San Jose, they would still be doing so. "When 51 Peg happened," Butler told me up at Lick, "we were on *Nightline* and in all the papers. Most astronomers can spend their whole professional careers without getting that kind of public recognition. When it happened to us, I figured that was our brush with fame. I had no idea it could get more intense than that. Now I know."

Chapter 6

WHAT IS A PLANET?

Just because Michel Mayor and Didier Queloz said they saw evidence of a planet orbiting 51 Pegasi and just because Geoff Marcy and Paul Butler said they saw planets orbiting 51 Peg and two other stars didn't make it so. Having your discovery on the cover of *Time* doesn't make it an established fact. All four astronomers knew that they were making dramatic claims; they knew of the tremendous public interest in other worlds and extraterrestrial life. But they also knew about Peter van de Kamp and Andrew Lyne, to say nothing of Giovanni Schiaparelli and Percival Lowell.

They knew, in short, that their observations would be examined minutely for mistakes, self-deception, carelessness, a failure to think of an obvious and perhaps much more banal explanation for what they thought they were seeing. If they hadn't questioned themselves before going public, they risked looking foolish at best. At worst, they could refuse to face the evidence against them should it appear. They would be scientific outlaws taking pot-shots at their critics from behind the barricades.

But they had thought things through. They had questioned themselves. And when they gave professional talks over the next few months at confer-ences and seminars—for Marcy and Butler, especially, were suddenly in great demand—they had little trouble getting through the question-and-answer period that always followed such talks.

I watched Paul Butler do it in May 1996 on a visit to Princeton Univer-sity. John Bahcall, the ex-president of the astronomical society who had per-suaded Andrew Lyne to issue his retraction in person, was in Butler's audience; so was Lyman Spitzer, the man who had helped demolish the planets-as-rare-commodities theory back in 1937. Jim Peebles was also pres-ent; he had been part of a Princeton group that predicted the characteristics of the cosmic microwave background radiation, the afterglow of the big bang, in 1964. "I find it rather incredible," Butler told me while eating cook-ies and sipping tea during the social hour just before the talk began, "that I should be here, lecturing before the gods of astronomy."

These particular gods weren't especially knowledgeable about planets. Princeton's astronomy faculty is celebrated for its work in cosmology, the study of the birth and evolution and structure of the cosmos as a whole. Such insignificant nearby objects as planets aren't much discussed.

Moreover, the Princetonians tend to be theorists rather than observers. They work with equations rather than telescopes, trying to uncover the un-derlying principles that govern the universe. It's a general rule in science that no theory can be accepted until it's confirmed by observation. Einstein's theory of relativity, for example, was just an ingenious exercise in higher mathematics until Sir Arthur Eddington showed that a beam of starlight passing close to the Sun would be bent slightly by precisely the amount Ein-stein had predicted it would. At Princeton, however, according to astronom-ical folklore, the general rule is turned on its head: No observation is accepted until it is confirmed by theory.

Despite their relative ignorance about planets, however, the Princetoni-ans' intuitive grasp of the basic issues allowed them to ask Butler plenty of challenging questions. At first these went to the issue of whether there were bodies orbiting around 51 Peg, 47 UMa, and 70 Vir at all. "How do you know you're not just seeing pulsations?" asked someone from the audience after Butler was finished. "That's a good question," replied Butler. "Because the planet around 51 Pegasi seemed so bizarre, people naturally suggested that something peculiar might be going on with the star itself. But in fact it's very much like the Sun, although at eight gigayears it's somewhat older [that is to say, 51 Peg is eight billion years old, to the Sun's five]. And Sun-like stars don't show such pulsations."

Still, he, Marcy, Mayor, and Queloz had had to consider the possibility. They would have been crazy not to. Plenty of stars do pulsate, waxing and waning in brightness in a regular cycle that lasts a few days or a few weeks. What is actually happening is that the stars are changing in size—swelling, then shrinking, over and over, almost like a beating heart—and, naturally enough, when they're bigger, there is more surface shining. The star gets brighter.

As the star expands, its surface moves outward, toward the observer. Even though the star as a whole isn't moving, its surface is. The light waves are compressed, their wavelengths are closer together, their frequency rises, and they get bluer. When the star shrinks, the opposite happens; the light is red-shifted.

How do you tell the difference between a star moving toward and away from you—a star being tugged by a planet—and a star whose *surface* is moving toward and away from you? This is what the Princetonians wanted to know. And the answer was simple: If the star is red-shifting and blue-shifting without brightening or dimming, there's no pulsation going on—or at least not the kind of pulsation that anyone has ever seen. Others had watched these same stars very carefully, said Butler, and had seen no variations in brightness within the limits of the observations.

"Could you be seeing the effects of star spots?" came the next question. This was another obvious, though subtler, way that Marcy and Butler could be fooling themselves. Because a star is rotating, its light is actually both red-shifted and blue-shifted at once, all the time, whether or not it has planets or pulsations. Light shining from the edge of the star that is rotating toward Earth is slightly blue-shifted; light from the other edge is slightly red-shifted (and light that is right in the center, from the part of the star that is moving directly across the field of view, isn't shifted at all).

The result is that each spectral line, both red-shifted and blue-shifted at once, gets thicker, and the faster the star is rotating, the thicker each line will be. In fact, you can deduce a star's rotation rate by seeing how thick its spectral lines are.

Just like the Sun, though, stars have dark spots across their surfaces, areas of cooler gas that mark the locations of magnetic storms. Some stars have especially large spots whose presence can trick unwary observers into thinking the star itself is wobbling.

What happens is that the approaching edge of the star will be darker than average as a spot rotates into view. Overall, therefore, there is less blue-shifted light than there would normally be, and spectral lines will visibly thicken only in the redward direction. As the spot rotates out of view on the other side, the red-shifted light will darken in turn; the spectral lines will thicken only blueward. Just as an advertising sign can simulate motion by turning lights on and off in a regular rhythm, so can a star appear to wobble when it is really standing still.

But Marcy and Butler are wise to this trap, too. The four-day wobble of 51 Peg, if it was caused by star spots, would mean that the star itself is rotating every four days. The thickness of its spectral lines prove, though, that it is in fact rotating every thirty days. That deduction is confirmed by the fact that 51 Peg shines with a constant brightness. Astronomers have learned through observations of thousands of stars that fast rotaters tend to have lots of changes in brightness. A quiet star like 51 Peg is almost certainly rotating slowly.

That made sense to the Princetonians. The objects really did appear to be there. Now the questions centered on how big the planets actually were. How did Marcy and Butler know they were seeing the orbits edge-on, the way you'd see a dinner plate if your vantage point were the edge of a table? This was an extremely important point. Jupiter makes the sun wobble by twelve meters per second, but only if you're looking edge-on. If instead you were looking down from above, so the planet's orbit looked circular, all of the pulling would be from side to side, across the line of sight. The Sun wouldn't be approaching you or receding from you at all, and its light wouldn't be either red-shifted or blue-shifted.

If you looked at the solar system from an angle of forty-five degrees, on the other hand, you'd see about half of Jupiter's influence. Half of the tugging would be toward and away from you, half would be back and forth across the line of sight. You'd only be able to detect the former. Yet if you made the naive assumption that you were seeing all the motion there was, you'd think you were seeing a planet significantly less massive than Jupiter.

So maybe 51 Peg B wasn't really as small as it looked. Maybe it was huge—a brown dwarf rather than a planet—and it only looked planet size because much of its influence on the star was in a direction that rendered it invisible. Marcy and Butler (and Mayor and Queloz) recognized this, of course. They were always careful to say that 51 Peg's planet, for example, was equal to a half-Jupiter in mass *at a minimum*. It could in principle be much bigger.

The observers didn't think it was bigger, though, and Butler explained why. Theorists believe that solar systems form as a unit, stars and planets alike, out of a single rotating cloud of gas. Both the star's rotation and the planets' orbits are motions left over from that original rotation. If the star's rotation is seen edge-on, the planets' orbits are also being seen edge-on. Thanks to the broadening of 51 Peg's spectral lines, the astronomers are convinced they're seeing the star's rotation edge-on. Ergo, the planet is edge-on as well, and none of its effect is being hidden. "We've done that for all of our stars," said Butler. "They're all consistent with these things not being pole-on."

The next question, from John Bahcall, was an easy one. How did they know their desmearing software really worked? Maybe the apparent motion of 51 Peg, 47 UMa, and 70 Vir were due to mistakes in the computer code. But if that was true, answered Butler, they should see planets around all stars, and most of the stars in their observing program didn't appear to wobble at all.

The dearth of planet experts at Princeton probably explains why Butler didn't have to confront the one serious scientific controversy that had surrounded his and Marcy's discoveries right from the beginning. Within min-

utes after Geoff Marcy had finished his press conference at the American Astronomical Society meeting in San Antonio, one small faction of astronomers had raised an important question that had not yet been answered
to its satisfaction. Okay, this group said. We admit you have good evidence
that there are objects orbiting these stars. But what makes you so sure they're
planets?

The straightforward answer is their mass. The planet around 51 Peg is
halfway between Jupiter and Saturn in size. The planet around 70 Vir is
about six and a half times as massive as Jupiter. (It's not physically any bigger, however; once a planet is as large as Jupiter, any additional mass just
compresses everything down. The planet gets denser but not any bigger.) According to theorists, the smallest brown dwarfs—those objects that are more
massive than planets but less massive than stars—should be about ten
Jupiters, at a minimum.

Mass, however, can be a highly unreliable guide to classifying heavenly
objects. Pluto is classified as a planet, for example, but it's considerably
smaller and less massive than Earth's Moon; than the four of Jupiter's moons
that Galileo discovered; than Saturn's moon, Titan; and than Neptune's
moon, Triton. But Pluto doesn't orbit another planet; it orbits the Sun directly, and it even has its own moon, Charon, discovered in 1978. That must
be the difference.

Or maybe not. At least one asteroid, Gaspra, orbits the Sun as well and
has a tiny moon orbiting it, too, discovered in a close-up photo taken by the
Galileo space probe. Nobody calls Gaspra a planet. In fact, some astronomers argue that Pluto really isn't a planet, either. Out beyond Neptune,
the solar system is surrounded by an enormous ring of what are basically
chunks of ice. (It's called the Kuiper belt or the Kuiper disk, after the Dutch
astronomer Gerard Kuiper, who proposed its existence.) Every so often one
of these objects is knocked out of the ring and falls Sun-ward on a long,
looping trajectory. The chunk heats up, spews out gas, dust, and a tail, and
becomes a comet.

Pluto, too, is a chunk of ice. Pluto lies right in the Kuiper disk. Isn't it reasonable to classify Pluto as the largest of the Kuiper disk objects—an over-
grown comet that hasn't fallen into the Sun—rather than a planet? Several

of the moons of Uranus and Neptune are also large, icy bodies quite similar to Pluto. Presumably, they're large Kuiper disk objects that were captured long ago by the gravity of the giant outer planets. A bit farther out there may well be Pluto-size objects that are too distant and too dim to have been discovered yet. Should Pluto be classified as a planet while these other objects aren't, simply because it happens to have ventured close enough to be seen but not to be captured by Neptune or Uranus?

In light of a more modern understanding of what Pluto really is, some astronomers now say it should simply be reclassified. Either stop calling Pluto a planet, they say, or accept the idea that the solar system probably has hundreds or even thousands of planets, slightly smaller versions of Pluto, wafting about unseen in the Kuiper belt.

Downgrading a planet, though, is not so easy, as astronomer Larry Esposito of the University of Colorado discovered in 1995. "I was being interviewed about Pluto by USA Today," he told me in a telephone interview, "and I just casually let it drop that it isn't certain that Pluto is a planet. This is not a controversial statement to planetary astronomers, but the public was very upset. For some reason people are really attached to the idea that there are nine planets."

Alan Boss, a planet formation theorist at the Carnegie Institution of Washington (a private research institution that functions essentially as a university with just a few graduate students, something like Princeton's Institute for Advanced Study), agrees that Pluto is a problem, though he doesn't think it has to be reclassified. It's fine for schoolkids to learn that Pluto is a planet, he says. It has always been classified that way, and you can't rewrite history.

"But one of the real eye-openers you get in grad school," he continues, "is that a lot of the things you learn in textbooks aren't really true." Just because humans come up with a particular scheme for classifying solar system objects doesn't mean that Nature will cooperate. Still, Boss believes, "it's really only the people who study the outer solar system in depth who need to know that Pluto is a transitional object."

In the same way, he argues, these new objects found by Mayor, Marcy, and their collaborators may start to blur the distinctions between true planets and brown dwarfs. Boss himself had declared a decade ago that ten Jupiters was the minimum mass for a brown dwarf, but he says there was always a factor-of-two uncertainty in his numbers. It could be twenty Jupiters. It could be five. Similarly, the maximum size a planet can attain was calculated by Doug Lin at about one Jupiter mass. (Lin is the same University of California, Santa Cruz, theorist who phoned Geoff Marcy as he came down from the mountain, right after the initial discovery of 51 Peg B.) But again, the numbers have plenty of inherent uncertainty.

In fact, the uncertainty in both numbers is great enough that the two kinds of objects could overlap. The most massive planet could actually be bigger than the least massive brown dwarf. If that's true, then there would be no easy way to distinguish one from another; big planets and brown dwarfs are made from essentially the same raw materials and would look almost identical if you stumbled on one or the other during a visit to an alien star.

Yet it makes an enormous difference whether the objects Marcy, Butler, Mayor, and Queloz are finding are big planets or small brown dwarfs. That's because, according to prevailing astronomical theory, planets should be born in litters, like pigs or puppies. Brown dwarfs should be born alone. If these are planets, there could be smaller planets out there, including Earths—if not in these particular solar systems, then at least in a respectable fraction of the systems yet to be discovered. If they are brown dwarfs, on the other hand, the case for the existence of other Earths won't have been advanced at all.

Both planets and brown dwarfs—and stars as well—get their start in the same way: in giant clouds of cold gas and dust, weighing as much as a million Suns or more, that float between the stars. The clouds, like the universe as a whole, are made mostly of hydrogen, plus a little helium. They're also contaminated with tiny amounts of impurities like oxygen, carbon, nitrogen, silicon, and iron—a mix of chemical elements cooked in the furnaces of an earlier generation of stars, then released into the cosmos as these stars aged and died. Finally, the clouds are flavored with an even tinier admixture of more complex chemical compounds—carbon monoxide, formaldehyde, water, ammonia, even alcohol—made from atoms that have drifted together and latched onto one another over the eons. When enough of these complex molecules stick together, they graduate into dust.

Gravity wants to make these clouds fall in on themselves, but even at temperatures of only a few tens of degrees above absolute zero—which itself is minus 459 degrees Fahrenheit—they are too warm. Their meager heat exerts just enough pressure to keep them inflated. The clouds balance on the edge of collapse.

Not forever, though. Just a tiny nudge—from a close encounter with another cloud or a powerful explosion from a dying star—can be enough to push a small region, a cloud within the cloud, over the edge. At the outset its density is almost incomprehensibly low: only .000000000001 gram of matter in the average cubic meter of space. If the cloud were oxygen instead of hydrogen, you'd need to inhale five hundred thousand cubic miles' worth to get a single breath of air. (I owe this dramatic and elegant analogy to my colleague Marcia Bartusiak, who created it for her 1986 book *Thursday's Universe.*)

As it falls together, the cloud spins faster and faster, conserving its angular momentum by recasting the slow, majestic spin of an object a light-year or more across as a tight, rapid pirouette. The density climbs from .000000000001 gram of matter per cubic meter to .000001 to one to a million grams per cubic meter—the density of water—and far beyond. Under ever-increasing pressure, as density rises, the gas heats up. It reaches tens of degrees above zero, then hundreds, then thousands. Finally, when it reaches ten million degrees or so, the pressure is high enough to force the nuclei of hydrogen atoms to merge with one another. Nuclear fusion, the same reaction that powers a hydrogen bomb, begins. A star begins to shine.

Or two stars, or three, depending on whether the mini-cloud stayed intact as it collapsed or broke apart to form distinct centers of collapse. The latter is evidently more common, since about 60 percent of the stars in the Milky Way are in double- or triple- or even quadruple-star systems. Or—and here, finally, is where brown dwarfs come into the picture—the mini-cloud breaks apart, but only one piece of it is massive enough to generate sufficient pressure in its core to burst into fusion. The other fragment (or fragments) falls a little short. It collapses but never gets hot enough to shine. It merely glows dully with the heat of internal pressure alone or perhaps with a low-grade type of hydrogen fusion. This failed star is not even a dim red, like the smallest, coolest stars, but shines only with dull infrared light. It is figuratively, though not literally, brown. And it is a dwarf object at most eighty times as massive as Jupiter and at the least— Well, that's the question.

If the cloud stays intact, though, without casting off brown dwarfs or extra stars, it will flatten as it spins, like a lump of watery clay on a potter's wheel. Only the very center of it will be dense enough to begin fusion. The rest remains a mix of gas and dust, swirling around a newborn central star.

This isn't just a vague theoretical notion anymore, nor is it something that exists only in the virtual-reality universe of a supercomputer simulation. Observers have actually seen flat disks of dusty gas encircling very young stars. In a recent survey of one hundred or so of these young stars, about half showed evidence of accompanying dust disks.

At this point, however, fact must yield to theory, and Alan Boss yields to his Carnegie Institution colleague George Wetherill, widely acknowledged to be the intellectual father of modern planet formation theory.

As he stood to greet me in his office at the Carnegie, on the northern outskirts of Washington, Wetherill certainly looked patriarchal. He is short, white-haired, gnarled. He looks like nothing so much as an old sailor. Wetherill was first exposed to planet formation theory in a series of lectures back in the late 1940s, before it really deserved the formal name of "theory." He was an undergraduate at the University of Chicago at the time and remembers being frustrated that the ideas floating around then were always rather vague. "I could never quite understand what they were talking about," he admitted. He ended up going into nuclear physics and then, because he didn't want to work on bombs, got involved with a project at the Carnegie Institution using radioactive decay to measure the ages of rocks. That eventually led to a professorship in geology at U.C.L.A. (though he'd never taken a geology course), and that in turn led him to an interest in meteorites, and finally back to the question that had frustrated him so long before: How did the solar system form?

By then the science had advanced considerably, thanks largely to the work of a Soviet astronomer, Viktor Safronov, during the 1960s. The work was meatier and much more detailed and analytical than the vague scenarios Wetherill had originally been exposed to. It was, in short, real science. This was in the early 1970s. Wetherill has been working on planet formation

ever since—the second field of science in which he has worked without having had any formal training.

The way things probably happened in the early solar system, he explained, was that the gas and dust disk surrounding the newborn star first
had to decide whether it would survive or whether it was going to fall into
the central star. If the star was massive enough—one hundred solar masses,
say—its gravity would simply suck everything in and consume the disk.

A star like the Sun didn't have enough gravity to do this, so the disk stayed
intact, and over time its dust would begin to clump together, forming rocky
chunks. Close in to the Sun, it was too hot for anything but rocky minerals
like calcium, magnesium, olivine, and pyridoxine to condense without evap-
orating again. Everything else was driven away by the heat and by the young
Sun's powerful solar wind. Only farther out could the lighter gases, including water vapor, start to form solid lumps (solid water, of course, simply
meaning ice). "That's out around Jupiter somewhere," Wetherill said.

As the disk gradually cooled, the chunks stuck together to form boulders,
and the boulders stuck together to form mountains about a kilometer across,
more or less—about half a mile. Eventually all of this solid stuff began to
settle in the central plane of the disk. The solids formed a hamburger of
kilometer-size rocks surrounded by a bun of gas.

At this size, the rocks were large enough to be governed more by gravity
than by the sticky chemical forces that originally formed them. They grew
larger and larger, and over a period of perhaps ten thousand to one hundred
thousand years there got to be hundreds of objects ranging in size from the
Moon to Mars.

These objects were the raw materials of the planets. They orbited around
the Sun, smashing into each other, occasionally sticking together, sometimes breaking apart again, flinging each other into new orbits, smashing
once more. On balance, they tended to grow fewer and larger as time went
on. It took perhaps another ten million years before the planets reached their
present sizes, although, says Wetherill, "there were significant numbers of
rather large bodies, including Mars-size objects, which were loose cannons
flying around in this swarm." According to the prevailing theory, such a
loose cannon—or cannonball, really—presumably smacked into the young

Earth, stripping off a giant blob of molten rock that solidified into the Moon.

Out around Jupiter and beyond, though, where it was cool, there was still a lot of gas lingering. So the rocky planets that formed out there—the presumption is that they weighed in at about ten times the mass of Earth—began vacuuming up this excess gas with their gravity. In about a million years, perhaps, you'd have a planet consisting of a rocky core surrounded by a thick, hydrogen-rich atmosphere thousands of miles thick. You'd have Jupiter or Saturn. Plenty of details have yet to be worked out in this scheme, admits Wetherill, but most astronomers think it's more or less correct.

This is why 51 Peg B came as such a shock to astronomers: According to standard planet formation theory, it shouldn't exist. The planet is bigger than Saturn, so it almost certainly has to be made mostly of gas. But it's much too close to its star to be anything but rock. Yet according to theory there can't be enough rock so close to a star to make such a big planet.

The astronomers who discovered 51 Peg B and the ones who confirmed its existence were not unaware of this problem. It is partly why they never bothered to look for an object like this one and found it only by accident. Why look for something theory says can't be there?

Once it was found, however, somebody had to deal with it. The planet was like a drunken relative at a wedding—unpleasant to look at but impossible to ignore. Most of the time, theory has to be confirmed by observation. Perhaps this was a case where the Princeton approach was appropriate. The lack of any reasonable explanation for how such a thing could exist might cast doubt on the observations themselves, solid though they seemed to be.

More typically, though, when confronted by what appears to be an impossible fact, theorists will simply readjust the parameters of the possible by coming up with a new theory. Before anyone would even consider disbelieving in the existence of 51 Peg B, theorists would do everything they could to reconcile existing planetary theory with this new, unexpected, and bizarre phenomenon.

The reconciliation might well have taken months or years but for the impossibility that was the planet around 51 Pegasi, it took less than a week. The man responsible was Douglas Lin. I finally caught up with Lin in his office

among the redwoods in Santa Cruz in September 1996, about a year after
51 Peg's planet had first been found and nine months after Marcy and Butler's triumphant press conference at the American Astronomical Society
meeting.

Lin is an athletic, muscular Asian-American man in his early forties. He
moves and speaks with strong self-assurance—a little too much for some
people's taste, evidently: Some colleagues consider Lin something of a loose
cannon. In their view, he seems to have a new theory for every occasion,
which can be the mark of recklessness unless, as in Lin's case, you often turn
out to be right. When he talks, he leans forward in his chair and looks at you
with an utterly focused intensity—tempered, fortunately for the listener, by a
gift for comprehensible analogy and a self-deprecating sense of humor.

"Was I surprised when I found out about the planet around 51 Pegasi?"
he asked rhetorically. "Well, I was surprised that they found it, yes. But I
wasn't surprised that it existed." Back in 1982, it turns out, Lin had written a
paper based in part on earlier work by the theorists Scott Tremaine and Peter Goldreich, proposing that giant planets shouldn't necessarily stay put
once they had formed. Under the right conditions, he argued, they should
spiral in toward the central star like a crippled satellite falling to Earth, eventually crashing and vaporizing in an incandescent flash.

Lin's point was that in a disk that happened to be particularly massive, a
planet like Jupiter would suck in all the gas it could reach, but that there
would still be lots left over. The giant planet would be orbiting in a clear
zone, but plenty of gas would remain, both between the planet and the star,
and out beyond the planet. While this gas would be too far away to be
sucked in, it would still pull on the planet and the planet would pull back—
and the net effect of all this pulling, it turns out, is that the planet would be
forced, inexorably, toward the star.

Lin's ideas received little attention at the time, since it seemed to be a
purely theoretical exercise. Jupiter and Saturn had obviously not fallen into
the Sun and weren't about to. In light of 51 Peg B, however, it all suddenly
made sense—or almost, anyway. "I can't tell a lie," Lin continued. "I'd said
planets would migrate inward. But I didn't expect them to *stop*." He had to
figure out a theoretically honest way that 51 Peg B could come within an as-

tronomical hairsbreadth of plunging into its star and then, at the last minute, not do so. Lin and his collaborators, Derek Richardson and Peter Bodenheimer, reasoned that something had to stop the disk of gas and dust from pushing the planet inward.

What they realized was that as a giant planet spirals closer and closer to its star, it should begin raising tides on the star's surface—for a star's gaseous surface is just as susceptible to an orbiting object's tidal pull as the Earth's oceans are to the pull of the Moon. These tides transfer gravitational energy from the star to the planet. The star's rotation rate slows down. At the same time, the gravitational energy that the planet robs from the star makes the planet's orbital speed increase, as though it had suddenly been equipped with a booster rocket. At a certain orbital distance, the extra speed just compensates for the forces of the gas cloud, and the planet settles into orbit.

Most of Lin's colleagues find this idea persuasive. "I think Doug is right about planets moving in," says Alan Boss, "and I think that's the most plausible explanation for 51 Peg. The thing comes screaming down and hits this 'wall' [of energy] and just stops—bang!" The fact of 51 Peg's planet was clearly much more palatable once Lin had come up with a theory to explain it.

If the idea of a Jupiter careening through the inner solar system sounds apocalyptic, it should. A direct hit on an Earth or a Mars or a Venus or a Mercury would be quite possible, and even a million-mile miss would be enough to knock the other planets out of their orbits as it went by. They'd be thrown by the spiraling giant's gravity, either into the Sun or out toward interstellar space. Based on this idea, you might assume that the chances of finding habitable planets in the 51 Peg system are negligible.

Not according to Lin, however. He argues that big planets could form and spiral in over and over. Our Jupiter might be the tenth Jupiter that formed. The rest could all have fallen into the Sun long ago, destroying nine previous generations of Earths. Only the last time was the lingering gas sparse enough to allow Jupiter to stay in place and the final Earth to survive. "I predict that around 51 Peg we'll find a whole bunch of planets farther out."

In response to this idea, Boss says, "Nothing can be ruled out, of course.

But planet formation is a complicated enough process without trying to make it happen twice. I would be astounded to find a retinue of terrestrial planets sitting out there." So would most other theorists—but nothing short of observations will ultimately settle the question.

What 51 Peg B does prove, however, is that it's possible for solar systems to exist that are entirely unlike our own. Even in the mainstream planet-formation scenarios that George Wetherill favors, computer models rarely come up with the same solar system twice. Sometimes you get an object in the position of Venus that has twice the mass of the object in the position of Earth. Sometimes you get four bodies in the inner solar system, sometimes five, sometimes three. Jupiters and Saturns can form at places different from those they occupy in our planetary system. Even an oddball like 51 Peg B didn't throw him too badly when he first heard about it. "I was slightly surprised, but not overwhelmingly. Within this general scenario, I think there are all kinds of planetary systems that can be formed. The way I think it'll end up," he continued, "is not only that planetary systems will look different but that probably there will be more than one way of making them."

Alan Boss says pretty much the same thing. Until now, we knew what our solar system was like, so all the theorists had to do was reproduce this one example. They were *post*dicting rather than *pre*dicting. Now they're in the theoretical danger zone because observers are finding actual systems to test the theories. "It's a much more honest situation for everyone involved," he says.

That still doesn't answer the question of whether these new objects really are planets, formed the way Jupiter and Saturn formed, or whether they're actually brown dwarfs—failed stars whose presence suggests nothing at all about the frequency of planets around Sun-like stars, and therefore nothing about the probability of finding life. It's difficult, as Boss noted, to distinguish between a large planet and a small brown dwarf based simply on its mass or composition.

Theorists do expect, however, that the two should be quite different in the shape of their orbits. Johannes Kepler realized early in the 1600s that the planets' orbits around the Sun do not trace out perfect circles but ellipses—squashed, elongated circles. The degree of departure from circularity is technically known as eccentricity.

The orbits of planets in our system aren't elongated all that much—just enough to throw off calculations that fail to take that data into account, like Kepler's attempts to predict the position of Mars or Andrew Lyne's effort to see if planets were orbiting his pulsar. In fact, astronomers routinely refer to the planets' orbits as circular, leaving the word "nearly" unspoken, because other objects have orbits that are so much more eccentric.

Double stars, for example, tend to revolve around each other in highly eccentric orbits because they form directly from collapsing clouds that are asymmetric to start with. But planets go through an intermediate step that tends to civilize them. The disk of gas and dust from which they form is thick and hot, the astrophysical equivalent of sludge. Any eccentricity in the original cloud is spread around, evened out.

By the time a solar system is fully formed, the surviving planets are moving in nearly, though not quite, perfect circles. A perfect circle has an eccentricity of zero. A parabola—an orbit so elongated that an object traveling it would have to go an infinite distance before circling back—has an eccentricity of one. The Earth's eccentricity is .02. Jupiter's is .05, Neptune's .001. Double-star systems, by contrast, generally have eccentricities of .1 and higher. Brown dwarf eccentricities should presumably fall into the same range.

70 Virginis has an eccentricity of .38.

That glaring fact hadn't escaped Marcy's and Butler's attention. An eccentric orbit is, as the name implies, off-center. The planet around 70 Virginis comes closer to the star during one half of its orbit than it does during the other half. During this close approach it zips around a bit faster. The acceleration was inescapably evident in the shape of the curve that snaked across Paul Butler's computer screen.

The eccentricity of 70 Vir B didn't escape the listeners at Marcy's press conference at the AAS meeting in San Antonio, either. Alan Boss was in the audience; immediately after Geoff Marcy finished, Boss buttonholed him in the hallway. How, Boss asked, can you call 70 Vir a planet when it has a mass more than six times that of Jupiter and such a high eccentricity? 51 Pegasi B has a very low eccentricity, but that could be because it's so close to its star. The tidal pull between planet and star would not only slow the star's rotation but would circularize an initially eccentric orbit. 47 UMa was the only object with a reasonable orbit and low eccentricity, but maybe *it* was the oddball—a brown dwarf with a circular orbit.

Marcy and Butler had thought about this already, as they had thought about every other objection anyone was likely to raise. They couldn't prove the three planets weren't brown dwarfs, but they did have an argument against the proposition. It was tough to follow Marcy's reasoning while standing there in the hallway, so I asked them to explain it—slowly—when the crowds had dispersed.

"Okay," said Marcy. "First of all, there's no question that 70 Vir B has high eccentricity, and there's also no question that this is a surprise, considering what the theorists believe about how planets form. We acknowledge that." But 70 Vir B is only 6.6 times as massive as Jupiter. If brown dwarfs really exist, they should range in size all the way up from a few times as massive as Jupiter to 80 times as massive (any more mass and they'd turn into full-fledged stars).

So where are all the objects bigger than 6.6 and smaller than 80? "If these really are brown dwarfs and if you have a 6.6 Jupiter mass object," Butler interjected, "then you'd better have a 20 and 30 and 50 and 60. There had better be a whole array of these things." Not only should they be there, but you should see them. They're more massive than 70 Vir, so you should have seen them before 70 Vir—if they were there. Yet while astronomers have been looking for these larger brown dwarfs for years, nobody has seen them with the exception of Dave Latham's object, HD114762. "It's a desert out there," said Marcy.

That desert, they argue, means that 70 Vir and the rest of the new objects are not brown dwarfs but something else. "They aren't exactly what you think of as planets, either," admitted Butler, "and, well, there really isn't a word. Perhaps if we were a race that roamed between the stars, we would have as many different words for 'planet' as the Eskimos have for 'snow.' In the case of these big guys, we're now calling them 'eccentric planets,' to distinguish them from regular planets."

"Sure they're calling them something like that," said Alan Boss when I spoke to him in Washington. "They want to find planets. There's more glory in finding planets than in finding brown dwarfs. That's my cynical pronouncement."

When I reported this reaction to Butler, he said, exasperated, "I think it's going to take these old-timers—they're not old in age but in mind-set—a long time to change. The theorists have been concocting solar systems for years and years and years. And invariably, with few exceptions, they end up explaining why it is that all solar systems should look more or less like our own. Well, of the first three systems we've found, two of them look radically different from that."

Only three systems, though, out of sixty stars Marcy and Butler had analyzed by January 1996, might also suggest that planets are extraordinarily rare. Back in the 1960s, Frank Drake and the scientists who gathered to hammer out the particulars of the Drake equation figured that maybe 20 percent of all Sun-like stars would have planets. More recently, astronomers had seen dusty disks around 50 percent of young stars. Now Marcy and Butler were reporting planets orbiting just 5 percent of the stars they'd observed.

The catch, of course, was that their survey, though more sensitive than any other, was not sensitive enough to say anything meaningful yet about the overall prevalence of planetary systems. Any statements about what 70 Vir B and 47 UMa B and 51 Peg B really were and how they had formed and what they implied about life in the universe were absurdly premature.

"We really don't have enough observations yet to decide whether all of these are planets or not," says Jack Lissauer, a planet formation theorist then at the State University of New York at Stony Brook and now at the NASA Ames Research Center. Lissauer agreed it was surprising that observers were finding 51 Peg Bs and the like before they'd found large brown dwarfs. It suggested to him that the big planets are indeed born differently from the way stars and brown dwarfs are born. On the other hand, their eccentricities suggested precisely the opposite conclusion. "That's why," he said, "I'm unwilling to come down on either side of the [planet versus brown dwarf] question yet." He was also unwilling to speculate on what Marcy and Butler's planets might portend for Earth-like planets, and for good reason. We're seeing these big planets because they're so massive and so close to their stars. We wouldn't be able to see our own solar system at all yet in an observing program like Marcy and Butler's. And whatever these new objects are telling us, they're saying very little, if anything, about the systems we're *not* seeing.

Lissauer knew—as Marcy and Butler and Alan Boss knew equally well—that the observers would have to look much longer and much harder than they already had before they could say for sure whether the planets they'd already found were sailing through space alone or were the largest objects in bizarre, unexpected types in the solar system—and just the tip of the iceberg. The observers would have to look at many more stars, too, than the couple of hundred they'd already checked, and do it long enough to find more conventional, slow-orbiting planets like Jupiter and Saturn. They'd have to push the limits of their instruments to find Uranuses and Neptunes, if these really existed.

Geoff Marcy and Paul Butler knew that after nearly nine years of work, and despite the dramatic discoveries they'd already made, they had only now reached the starting line.

1

Paul Butler, the junior member of the most successful planet-hunting team in history.

2

Geoff Marcy, who initiated the San Francisco State University planet search project; he and Paul Butler worked in vain for more than eight years before they began finding planets.

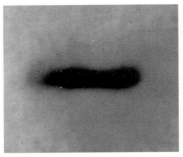

A disk of gas and dust surrounding a newborn star in the Orion Nebula. Within a few tens of millions of years, the dust may evolve into a solar system like—or perhaps very unlike—our own.

A readout from the Project Phoenix computer screen. Most of the dots are random radio noise, but the line slanting from left to right is a signal from the Pioneer 10 spacecraft. An alien signal might look something like this.

The Palomar Testbed Interferometer, seen from the catwalk of the old 200-inch Hale Telescope. The PTI, which combines two parallel beams of starlight to measure stellar positions precisely, is being used to perfect technology for a space-based instrument to be launched in 2003.

6

The nearly 14,000-foot summit of Mauna Kea, in Hawaii. The domes of the twin Keck telescopes are second from the left. Planet-hunters have brought their search to Keck I and Keck II, the world's most powerful telescopes; within a few years astronomers will combine the light from the Kecks in order to measure stellar positions—and infer the presence of planets—with unprecedented accuracy.

The Keck Interferometer (shown in an artist's visualization) will be the next step after the Palomar Testbed Interferometer, combining light from the two giant Keck telescopes with light from six smaller telescopes to hunt for other worlds.

7

10

Roger Angel, the University of Arizona astronomer whose ingenious approach to telescope design has revolutionized ground-based astronomy. He's now turning his formidable attention to space telescopes.

8

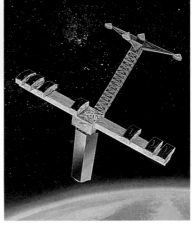

SIM, the Space Interferometry Mission, in an artist's rendering. When it goes into orbit in 2003, SIM will be able to find indirect evidence of Earth-size planets orbiting other stars.

9

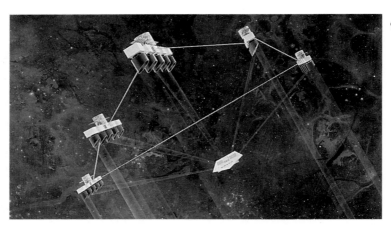

Planet Imager. No one can even guess when this super-interferometer might be built; it's currently beyond any imaginable technology. But NASA Administrator Daniel Goldin hopes a device like this might someday photograph a distant Earth-like planet in such detail that clouds, oceans, and mountains are visible.

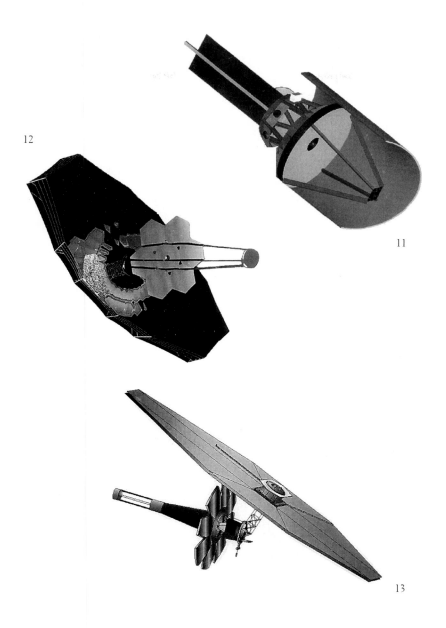

12

11

13

Three competing concepts for the Next Generation Space Telescope, a larger version of the Hubble; it is tentatively scheduled to start observing by the year 2007, in an orbit 1,000,000 miles from Earth.

14

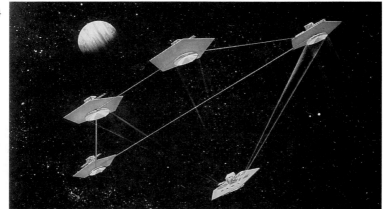

Planet Finder. A set of four free-flying telescopes, orbiting the Sun out in the vicinity of Jupiter beginning in 2015, will send their light to a fifth satellite; the combined light will form images so precise that NASA hopes not only to take pictures of distant Earths, but also to study their atmospheres for signs of living organisms.

A giant molecular cloud, part of the Eagle Nebula. The tiny projections may be fragments of the cloud that are collapsing to form new stars, and with them new solar systems.

15

16

A rock from Greenland; chemical evidence found inside suggests that life may have arisen on Earth at least 3.85 billion years ago.

17

Four views of Titan, Saturn's largest moon. Titan is massive enough to have an atmosphere, making it yet another place in the solar system where life could conceivably exist.

18

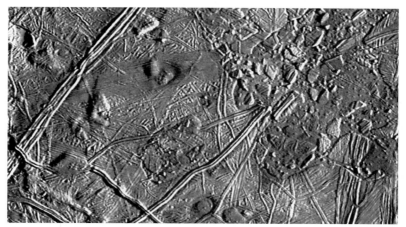

The frozen surface of Jupiter's moon Europa, as seen by the Galileo spacecraft. Cracks that crisscross the ice and chunks of ice that have floated and then refrozen in new positions show that there's almost certainly an ocean underneath, and with it the possibility of life.

19

Frank Drake, of the University of California, Santa Cruz, and the SETI Institute. Drake's project Ozma launched the modern search for extraterrestrial intelligence in 1961. His Drake equation has essentially become the foundation of a new branch of science, called astrobiology or bioastronomy.

The 140-foot radio telescope at Green Bank, West Virginia, where scientists from Project Phoenix are searching for signals from alien civilizations.

20

Chapter 7

SHARPER VISION

The summit of Mauna Kea, on the big island of Hawaii, is not a nice place to be. Gentle tropical breezes do not blow here. Palm trees do not sway. The daytime temperature in December is generally a few degrees below freezing, but at least this isn't the blizzard season. That comes in July. At nearly fourteen thousand feet, the air is only technically breathable. Atmospheric pressure, and therefore the amount of available oxygen, is less than half what it is at sea level. A brisk walk can leave you gasping for breath. Even standing still you can develop a terrific headache and nausea from oxygen deprivation, and if you remain up here for any length of time, you can develop full-blown altitude sickness, in which fluid accumulates in the lungs and brain. It can be fatal.

For that reason, people who plan to spend any time at the summit are supposed to spend at least twenty-four hours first at Hale Pohaku, a small

complex of buildings at the nine-thousand-foot level that includes dormitories and a dining hall, to give the body a chance to acclimate.

Paul Butler, Geoff Marcy, and Steve Vogt knew all this, but they were in
a hurry. So instead of twenty-four hours at Hale Pohaku, they spent fifteen
minutes—long enough to drop their bags in the dorm and grab a four-
wheel-drive sport-utility vehicle for the final few steep and treacherous miles
to the summit.

It is now December 1996, eleven months since Marcy and Butler stood
in a Berkeley office, two obscure astronomers staring in disbelief at a computer screen. Now they were celebrated as the most prolific planet-hunters
in history. From whenever in the dim past prehistoric humans first noted
that a handful of stars refused to stay fixed in place, until 1791 when
William Herschel discovered Uranus, only five planets were known to exist
beyond the Earth. In less than a year, Marcy and Butler had confirmed the
existence of 51 Peg, found 47 UMa and 70 Vir, and gone on to find three
more planets on their own and a fourth in collaboration with their friend
Bill Cochran.

Butler himself is not impressed. "Stamp collecting," he calls it. "We have
to start doing some science." Preposterous though that sounds, he's right.
The fact that planets exist at all, out there in the universe, is of course pro-
foundly significant and remarkable and satisfying. But if you're really going
to be serious about nailing down the first part of the Drake equation, you
need not a handful but a large sample of solar systems. A handful doesn't
give you enough information to estimate overall the number of planets in
the galaxy and to determine how many of them might be similar enough to
Earth so that life might arise there. It doesn't tell you whether our own solar
system is typical or so bizarre and abnormal that there's little hope of finding
anything like it. Making real progress through the Drake equation with what
they have so far is like trying to predict the outcome of a presidential elec-
tion based on the first seven votes.

Marcy and Butler need more planets, especially smaller ones. That is
why they came to Mauna Kea with Vogt, who has gone from unofficial to official member of their scientific team. The air here is some of the clearest
and driest and most stable on the planet. Thanks to trade winds that blow

steadily across the stratosphere, the skies above Mauna Kea have very little of the atmospheric turbulence that makes stars twinkle and drives astronomers insane. The "seeing" on Mauna Kea—the ability to discern the stars without distortion—is so good that to do any better you'd have to go out into space. That's why this is considered one of the world's best astronomical sites, why a dozen observatory domes dot the summit.

Among them is a rectangular one-story building, about 150 feet long, with not one but two domes bulging from it, one at each end. Inside the domes are the two largest and most powerful light-gathering machines in existence: the identical twin 394-inch-diameter Keck telescopes, built with private money and operated by a consortium that includes the University of California, Caltech, and the University of Hawaii. When Keck I was completed in 1992, it suddenly made the 200-inch telescope on Palomar Mountain in southern California—the world's largest at the time—look puny. Keck II was finished four years later. With a mirror thirty-three feet across, each of the Kecks can suck in four times as much light at one time as the 200-inch telescope at Palomar, and ten times as much as the 120-inch at Lick.

To get to the summit, the astronomers have to drive a road that winds uphill from Hale Pohaku in a series of sharp switchbacks, climbing about four thousand feet in only nine miles—an extraordinarily steep grade—and angling up through a landscape that changes almost immediately from spare to utterly barren. The low, scrubby bushes that struggle to live in the thin air and cold of the base camp are gone with just a few hundred feet more of altitude. After that, there's nothing at all covering the ground. The only colors visible are blacks and grays—volcanic ash, rocks, and mud. The first five miles of road are cinder, unprotected by guardrails despite occasional dropoffs of hundreds of feet; the last four are paved. The State of Hawaii began paving from the top and ran out of money.

The cinders are actually safer to drive on than the pavement when the road is iced over, as it frequently is, but the paved segment has guardrails. It's kind of a tradeoff. No astronomer has driven off the edge yet, although some have probably deserved to. The posted speed limit in the downhill direction is twenty-five miles per hour. A couple of Frenchmen decided to set a speed record and did it at sixty. They made it safely down in nine minutes flat. A visiting Japanese rock star, driving considerably slower, did go over the edge and was killed just above Hale Pohaku.

This is one of the last volcanic mountains that formed in Hawaii—or, more properly, that formed Hawaii; the islands are the exposed tops of volcanoes that rise from the ocean floor. As the Pacific tectonic plate moves slowly to the northwest in slow-motion continental drift, a stationary subsurface hot spot in the partially molten mantle deep underground stirs periodically. A new volcano appears, and eventually a new island. The hot spot dies off, the volcanic mountain moves gracefully off, and a new section of the Earth's crust slides into place, making room for a new volcano.

Mauna Kea is a recent volcano, but it no longer sits directly on top of the hot spot; it has moved north and west in its turn, and while it is still considered active, it is currently dormant. Nobody expects Mauna Kea to erupt again anytime soon—obviously, considering the hundreds of millions of dollars' worth of telescope that sits up here. The island hasn't moved all that far, though: The next mountain over, its summit clearly visible twenty-six miles away, across a cloud-filled valley, is Mauna Loa. Mauna Loa is almost as high as Mauna Kea but is still active. It last erupted in 1984, and geologists have no doubt that it will erupt again.

Finally, the gravel roadway turns to asphalt, the vehicle rounds a curve, and suddenly, there is the summit—a black landscape punctuated by the gleaming white of observatory domes. The Keck building is straight ahead, but while the three astronomers have plenty to do in the four hours until sunset—load their software, meet the telescope operator, make sure everything's functioning normally, go over the list of stars they want to observe one last time—they can't resist taking a short detour around the other domes.

They drive past the University of Hawaii's 88-inch, the Canada-France-Hawaii 144-inch (where Bruce Campbell and Gordon Walker worked on their own planet project), the 150-inch United Kingdom Infrared Telescope, and past domes that are still under construction for two more telescopes.

Now, finally, they drive up to the Keck building. They'll be working at Keck I during this run. The two telescopes are virtually identical—it's even

difficult to tell which is which when you're inside—but the electronic detectors the planet-hunters depend on are permanently installed on Keck I
(Keck II has a different set of detectors, useful for different sorts of astronomical problems). They grab their backpacks filled with notebooks and computer tapes and plenty of bottled drinks. It's not only airless up here but also
so dry that it's easy to get dehydrated, on top of everything else. Experienced
visitors keep a bottle of Gatorade handy at all times, taking little sips every
three or four minutes, through the night.

Before they go inside, they take a look straight up. It's completely overcast
down below; Mauna Loa looks like an island floating on a sea of white; part-
way up the mountain they had experienced this layer as a thick fog. They'd
been a little bit worried about running into a cow from one of the ranches
that straddle the road. Overhead, though, the clouds are thin and wispy. Not
great if you're looking at galaxies or quasars at the edge of the universe, but
no problem at all if you're looking at stars. "Looks good to me," says Butler.
"Let's do some science."

Before they settle in for the night, though, the astronomers step for a few
minutes into the dome itself, just on the other side of the wall from the control room. There is really no need for them to go in here. In the old days,
during the careers of astronomers who were senior scientists when Marcy
was a student, observers would spend hours on end sitting in the dark in the
prime focus cage, a sort of crow's nest high above the telescope's primary
mirror, sliding photographic plates in and out of their holders. Computer-
controlled telescope drives were unknown back then, and the astronomers
had to fine-tune the telescope's aim by hand to keep a particular star or
galaxy from blurring out as the Earth turned underneath them. Some astronomers would wear electrically heated suits, designed for jet fighter pilots,
to ward off the freezing cold of an open observatory dome. To hear them tell
it, the suits didn't work all that well.

Astronomers don't have to deal with any of this anymore. Every major
telescope now has a nice heated control room to sit in. In most observatories,
including Lick, this is basically just a closet, a tiny, cramped space with racks
of electronics spilling wires all over and computer monitors piled one on top
of another. The Keck control room is more like the bridge of the Starship

Enterprise. It's bigger than most living rooms. The electronics sit against one wall, neatly out of the way. Bisecting the room is a curving console, twenty feet long, with eight or ten monitors parked on top. You could easily fit a couple of La-Z-Boy recliners and a couch in the unused space.

The overall effect is one of sleek, uncluttered technology, appropriate for the premier telescope on Earth. Computers keep the Keck aimed and tracking whatever celestial object you might want to observe; they monitor the light entering electronic detectors; they whisk images off the detectors and onto storage tapes. Unless the astronomers have to adjust the spectrometer or do tests of some sort, they never really have to enter the dome at all.

Still, they do it, out of respect for their scientific forebears and out of respect for this gigantic, powerful instrument. They lead me through a door adjacent to the control room. It's freezing cold in here — even colder than it was outside. At most observatories the telescope operator opens up the dome an hour or two before the astronomers are going to observe, in order to let the temperatures equalize. If they didn't, the temperature differences between glass and air would set up turbulence over the mirror's surface, blurring the images of the stars. At the Keck they don't want to waste time, though, so the dome is air-conditioned even though the temperature outside the building is hovering around freezing. When the tall, thin doors in the curved ceiling finally slide open, the mirror will already match the night-time air.

The Keck dome is a huge, cavernous room with a hemispheric white ceiling like that of a planetarium and catwalks clinging to the circular walls at several levels. Stairs connect the catwalks to each other, but the astronomers know how to behave at fourteen thousand feet when you're not used to it: We take the elevator, which rises slowly, and then the doors open.

There, across dozens of feet of empty space, is the heart of the telescope: the largest mirror in the world, thirty-three feet across. And it's aimed right at us. We can see ourselves reflected in it, our images pretty much undistorted. A telescope mirror is curved to gather and focus light, but the amount of curvature is nowhere near as great as it is in a funhouse mirror. We also see easily that the mirror is made up of thirty-six separate hexagons of glass, fitted together to simulate a single, continuous slab. Making a mirror this size from one piece of glass would be extraordinarily difficult. Nobody has ever tried it, and not even the most ambitious mirror makers alive have expressed much interest.

Each hexagon is about six feet across and about two inches thick, and is held in place by three adjustable pistons, which are attached to the back of the glass at one end and to a giant Erector set scaffolding of metal beams — the mirror support structure — at the other. Every week or so a technician gets onto a long-armed cherry picker and rides up until he or she is inches from this wall of mirrors and sprays the reflective front surfaces with compressed carbon dioxide to blow away the film of dust that settles on the mirrors every night.

Every few months, on a rotating schedule, each hexagon of glass is pulled out entirely for refinishing. The telescope is pointed directly skyward; technicians get underneath and push the appointed segment up a few inches. Then an overhead crane descends from above and grabs the mirror with three claws, like the treasure-plucking game at an old-fashioned penny arcade, and removes it. The glass is stripped, then coated with a new layer of vaporized aluminum to make it as shiny as possible. The man who designed the system is no longer with the Keck. He's now designing roller coasters at Disney World.

Directly across from the mirror — right next to where we're standing — is a huge aluminum box, sitting on a platform that hangs off the catwalk. It's essentially a normal-size room within the dome; it even has a door, which is now closed and locked. This is Steve Vogt's latest creation: the high-resolution spectrograph he has spent the last five years building for the Keck. Armed with one of Marcy's and Butler's iodine cells, as it now is, this device, which Paul Butler describes as "incredibly gorgeous," will do what the Hamilton did, only better. The Keck telescope has about ten times the light-gathering power of the 120-inch Shane at Lick. The spectrograph is about twice as sensitive as the Hamilton.

Together they'll give Marcy and Butler twenty times the planet-searching capability they had in California; they'll be able to examine 20 stars in the time it took them to observe just one back at Lick. Their original survey included 120 stars, within a few tens of light-years of Earth; the new, Keck-based survey will cover 400 stars, out to a distance of two hundred light-years.

All of the observations, moreover, will be as good as the best Marcy and Butler could do in California. They'll be able to measure wobbles of just two or three meters per second, and perhaps a shade better. "We're going to look for ten years," says Marcy, "to survey for Jupiters and Saturns within five a.u. [that is, 450 million miles, the orbital distance of the actual Jupiter] of all these stars." They're going to put a microscope on these regions. This sensitivity will also let them spot Uranuses and Neptunes if they happen to orbit very close in, as 51 Peg B does.

This is precisely the sort of information that theorists like Jack Lissauer, Alan Boss, George Wetherill, and Douglas Lin need if they're going to begin understanding the general process of planet formation, as opposed to the specific process that formed our solar system. And it will be crucial in coming up with a final figure for F_p, the term in the Drake equation that describes the fraction of stars with planets.

Powerful as they are, though, the Kecks are just two telescopes; even if they could be devoted full-time to planet hunting—which, given the enormous range of phenomena studied by astronomers, they cannot—the local galactic neighborhood alone contains tens of thousands of stars. If astronomers are ever to solve the Drake equation, they will need more big telescopes.

It was thus a great relief to the astronomers who gathered in San Antonio in January 1996 that Geoff Marcy's dramatic announcement of 47 UMa B and 70 Vir B was followed later in the conference by a speech by NASA administrator Daniel Goldin. The agency was reorganizing its space science division, he said, into four programs that would reflect four broad themes: the relationship of Earth to the Sun; the exploration of the solar system; the structure and evolution of the universe; and, finally—and most interesting to those who cared about the question of life on other worlds—the Origins Program. Origins would be a search for, among other things, planetary systems around other stars, including Earth-like planets; for evidence of life on those worlds; and for the origins of life on Earth.

Unlike some proposals that have been floated about the space program—George Bush's presidential declaration that we would send humans to Mars within three decades comes to mind—Origins wasn't simply a vague list of appealing notions. Goldin named specific instruments that he planned to have built to further the search for life in the universe, and specific timetables for their construction. Among them would be the Next Generation Space Telescope (NGST), much bigger and more powerful than the Hubble, to be launched into space by 2007. "Finally," he said, "I have a dream that we'll someday build a telescope powerful enough to photograph the surface of an Earth-like planet orbiting a distant star, with enough resolution to distinguish clouds, continents, oceans."

This last proposal is generally considered pretty much a fantasy. It would mean a device more than a thousand miles across. High NASA officials, who would rather swallow poison than contradict the administrator publicly, will only say when pressed that Goldin's dream is "not impossible"—bureaucratic shorthand for "don't hold your breath." Even the Next Generation Space Telescope is extraordinarily ambitious and will be difficult to pull off, both technically and, just as important, financially. Congress is not writing blank checks for the space program anymore.

Origins does have two points that will give it an advantage in prying money from Congress. First, it addresses a set of questions that scientists themselves care deeply about. Second and even more to the point, any discovery that addresses the question of life on other worlds will inevitably capture the imagination of the bill-paying public—as Geoff Marcy's announcement a few hours earlier was already making clear, and as a meteorite from Mars, an ocean on Jupiter's moon Europa, and the perambulations of a Mars-roving vehicle named Sojourner would prove over the following eighteen months.

For almost three decades NASA had been hoping to recapture some of the overwhelming public support it had had during the Apollo moon landings of the 1960s. The agency had managed to do that occasionally, with the Viking landings on Mars in the mid-seventies, with the Voyager flybys of Neptune and Uranus in the late eighties, with Hubble observations of comet

Shoemaker-Levy smashing into Jupiter in 1994. But these events were transitory. Origins would be a sustained program with specific milestones and a long-term popular goal. Perhaps it would be the second coming of Apollo.

Daniel Goldin's job as NASA administrator is to set that sort of goal, to be a strategic thinker and a visionary, and then to act as a cheerleader and advocate for the agency. Goldin is especially good in the latter roles. If anyone can sell Origins to Congress and the public, he can. To make Origins actually work, however, it will take someone like Ed Weiler. Origins had already been under way informally when Goldin made his announcement, and when I caught up with Weiler in his office at NASA headquarters in Washington several weeks later, he had just been formally named its director. Before that, Weiler had been NASA's chief scientist for the Hubble space telescope, a title he would continue to hold.

Ed Weiler is a shorter than average man with medium-short blond hair. If you took the 1970s-era popular singer and frequent *Love Boat* guest star Paul Williams and gave him a haircut, the two would bear a passing resemblance. Weiler is among the least formal and least self-important high-level bureaucrats I've ever met. That may be partly because he started life as an astronomer rather than a government employee. But even among astronomers, Ed Weiler is something of an anomaly. Where some of his colleagues, present and past, might refer to the search for planets in terms of Copernicus and Kant, of the great quest for humanity to understand its place in the cosmos, Weiler—who is perfectly capable of such high-minded rhetoric—tends to prefer more accessible references. "The first time Wilma Flintstone looked up at the sky," he told me, trying to put his own spin on the public's fascination with Origins, "she probably said, 'Hey, Fred, come out and look at this. What are they?' "

Weiler knows from personal experience how important it is to keep the public excited about NASA programs and also how tough it is for the agency to live up to its own hype. It was Ed Weiler who had had to face reporters and explain back in 1990 why the Hubble Space Telescope couldn't focus properly. "I started working on Hubble *after* the mirror was polished," he was quick to establish. "But, still, it was an emotional roller coaster." At the time, Weiler was starting a family, and he said, "If you have babies, you can imagine what it was like getting up at two and four in the morning and then having to do a press conference the next day where they're asking you if you've stopped beating your wife. Which is what I did for about forty days in a row."

Since the 1993 shuttle mission that repaired the telescope—a feat no one believed the astronauts could really pull off, and one for which Weiler gets much of the credit—his job has been transformed. It used to be that he had to scrounge for Hubble pictures to show the public. Now he has to reject 95 percent of what astronomers give him. "When you have someone announcing the discovery of interstellar tin . . . it's interesting in its own way, but it's not going to get Dan Rather excited, if you know what I mean."

Planets around other worlds are something else entirely. Even Weiler was taken aback by the intense public reaction to Geoff Marcy's press conference—although he's pleased to take advantage of it. The cover of *Time* magazine featuring Marcy's and Butler's discoveries has become a standard part of the slide show he puts on whenever he gives a talk about Origins. Weiler bristled, though, at the suggestion that Origins was itself a reaction to the discovery of planets. "It was a coincidence, although I admit it was a wonderful coincidence for us."

In fact, there isn't much new about Origins except for the packaging (which is, admittedly, crucial). The program mostly just organizes a number of existing or already contemplated projects into a rational sequence that can be marketed to Congress and the public. The unsurprising truth is that NASA has been thinking about an organized search for other worlds for several years. In 1990 a workshop organized by NASA's Solar System Exploration Division had put together a document entitled "Toward Other Planetary Systems"—the TOPS report, in its NASA-fied acronymic form. The report, compiled mostly by astronomers from outside the agency, suggested that NASA fund a long list of planet-search projects, including observations from the ground and, ultimately, from space, above the Earth's obscuring atmosphere.

Then, in 1994, the agency organized another team of outsiders, called the ExNPS advisory group (in English: Exploration of Neighboring Planetary Systems). This led to a second report, which laid out essentially the same program in greater detail. At about the same time, NASA also funded a third study—known either as "HST and Beyond" or as the Dressler report, after its chair, Carnegie Observatories astronomer Alan Dressler, in which it asked astronomers to figure out what the logical successor to the Hubble Space Telescope might be and what it might be good for. All three reports had come to a tightly overlapping set of recommendations, and it was out of these that Origins was distilled. "You'll note that these recommendations came from the astronomical community," said Weiler, sensitive to the popular perception that NASA's main mission is to keep NASA in business. "In other words, this is not just Dan Goldin's or my crazy idea."

Weiler is also infused with the spirit of a space agency that has embraced new thinking—the idea promulgated by Goldin that NASA is going to do things "faster, better, cheaper," in a phrase that is repeated as a mantra around NASA. The Origins Program is being parceled out to several different NASA facilities, including the Jet Propulsion Laboratory (JPL) in Pasadena, California, and the Goddard Space Flight Center, in Greenbelt, Maryland. "One of my jobs," said Weiler, "is to make sure we don't go back to the old NASA way of doing things, with each center being its own little kingdom, so that if JPL needs a gyro they develop a JPL-model gyro and if Goddard needs one they develop a Goddard-model gyro. The not-invented-here mentality. We can't afford that."

Evidently not. NASA's budget overall is almost certainly going to be dropping—by 35 percent over the next six years alone, according to Weiler. Yet he and Goldin are talking about building a Next Generation Space Telescope (NGST) that will be at least 4 and probably closer to 8 meters in diameter (Hubble is 2.4 meters across) and sending it a million miles out into space (Hubble orbits at about 350 miles up). Hubble cost $2 billion. The NGST, they insist, must cost no more than $500 million.

This sounds insane on its face, but Weiler pointed out that technology has come a long way since Hubble was designed and built in the 1970s. Its fine-guidance sensors use 1960s-era photomultipliers. It stores data on old-fashioned magnetic tape. Compare the slow, dumb, bulky personal computers of the late 1970s with a modern Pentium-based workstation—thousands of times more powerful but no more expensive—and Weiler's optimism doesn't seem quite so wide-eyed.

Futuristic space telescopes aren't all there is to Origins; the program will also be funding research into how life arose on Earth; into how adaptable organisms are to extremely hot or cold or dark or airless environments, which will help set the boundaries of where astronomers should look for life on other worlds; and into what effects life might have on the mix of gases in a planet's atmosphere, which will help guide life-detection observations.

Origins will also take over the funding of projects already under way, which NASA has been supporting under the earlier TOPS and ExNPS proposals. Geoff Marcy and Paul Butler were partially funded by NASA money while working at Lick, and they're getting more for the Keck project. The Keck II telescope, in fact, was built partly with NASA money, with the search for planets explicitly in mind.

But it's the huge, wildly ambitious space-based projects, naturally, that have NASA managers and scientists most excited—although to hear project manager John Mather lay out his strategy for building the Next Generation Space Telescope, you'd think he was describing something about as inspiring as a new model pickup truck. He's mastered the utterly understated matter-of-fact NASA tone of voice—the "Houston, we have a problem" voice that astronaut James A. Lovell, Jr., used to tell ground controllers that Apollo 13 had just suffered an explosion halfway to the Moon.

Mather is a longtime NASA veteran, a physicist who proved himself as a project manager by running the Cosmic Background Explorer satellite program. In 1992, COBE discovered a phenomenon that astrophysicists had been looking for since the late 1960s: ripples in the radiation left over from the big bang itself. The man who received the public credit was George Smoot, a Berkeley astronomer who was responsible for the instrument that made the detection—but Mather was in charge of the whole satellite.

He oversaw COBE from his office at the Goddard Space Flight Center, just off the Washington Beltway, and it is from Goddard that Mather will be managing the NGST project. When we met at his office in October 1996, a year after he had taken over NGST, my first impression was that Mather could qualify as an Eagle Scout on demeanor alone. He is tall and skinny, with an utterly earnest, precise, and straightforward way of talking. I kept picturing him in tan shorts, wearing a sash covered with merit badges. When we went down to the Goddard cafeteria for lunch, Mather brought along a brown paper bag containing a sensible sandwich and a sensible salad. If your high school had an electronics club or a computer club or audiovisual club, Mather—or someone indistinguishable from Mather—was a member.

I asked him how he ended up in charge of the Next Generation Space Telescope, the bigger, better, cheaper version of the Hubble. Once COBE had completed its mission, he told me, he had to figure out what to do next. He'd already been thinking about the question of how you build a big telescope in space. Then, just about the time that Mayor and Queloz were announcing the existence of a planet orbiting 51 Peg, in the fall of 1995, Mather learned that he wasn't the only one thinking about it. "I got a phone call from Ed Weiler one day, asking me to take over the NGST project," he said.

It was a coincidence that this happened to be just the project he'd been thinking about, but it wasn't surprising that Weiler had tapped him. "The fact that I'd made one impossible project work already helped, I'm sure. Ed probably thought, 'Let's find someone who's crazy enough to try this.' Actually, he didn't ask me. He said, 'You're working on this, and I need your proposal by tomorrow.' So I wrote it up in a hurry—it was just an outline, only about three paragraphs, saying, 'This is what we want to do, and you should give us some money to do it.' He sent me enough to begin studying the problem, and I got started."

Ordinarily, the next question I would have asked someone in Mather's position would have been about what he and NASA hoped to accomplish with the NGST. In view of the cost limitations Dan Goldin had set for the project, though, I felt compelled to get something more basic settled first: Was this thing even possible?

"Well," answered Mather, "a lot of people don't think it will work, of course. And I agree with them—for now." The problem is that NGST will have to be radically different from any telescope ever built. That's because a telescope larger than the Hubble but built in a conventional way would be too large and too heavy to go into space on any existing launcher. The Next Generation Space Telescope, just about everyone agrees, will therefore have to be both lightweight and deployable—which is to say that it will have to go into space folded up in some way and then unfold in orbit.

The problem with that concept is that unfolding things in orbit is asking for trouble. "Whenever you hear about a big screwup in space," admitted Mather, "it almost always involves something mechanical—a stuck bolt that keeps you from deploying an antenna on the Galileo space probe, for example. But I feel that this attitude disables the thinking process, since it keeps you from looking at a whole class of potentially important ideas."

It's true that NASA doesn't have a good record at unfolding large objects in space, but the Department of Defense, by all accounts, does. The details of that work are classified, and John Mather doesn't have the security clearance to find out about them. What he can do, though, is have the DOD or its contractors build things for NASA. "They can say, 'Yeah, we can do that,' as long as they don't say *why* they know how to do it." It's pretty clear, he says, based on conversations with the Pentagon, that space deployment is technically feasible.

That doesn't make it cost effective, however. The DOD isn't especially well known for thrift. When the Air Force wants a new fighter, for example, it pays two companies to build competing versions of the same fighter, then compares their performance to decide which one to order. Clearly, NASA can't afford to do that, but it also can't afford unpleasant surprises. "The agency won't approve anything more than $500 million," said Mather, "and they won't approve cost overruns." If the project goes over budget, he believes, NASA will kill it without ceremony. So his plan is not to believe an instrument or a deployment scheme will work without proof. By the time he's actually starting to build the NGST in 2003, Mather doesn't want any unknowns.

How is that possible? "Well, maybe you test some of the components in the lab and test others in space. And we at NASA have to be on top of the technology and be very smart buyers." The agency will be doing its own development work, staying ahead of the contractors wherever possible and feeding them its internally generated ideas. Some outside contractors were worried at first that NASA would stay too far ahead and eventually decide to build the telescope itself. Mather insists that won't happen.

By now, a year after he'd gotten the call from Ed Weiler, Mather had assembled a staff of five or six people to work full-time on the Next Generation

Space Telescope, and he'd had input from hundreds more. During the preceding summer he had organized a series of workshops that included scientists and engineers from universities, government labs, aerospace companies like Lockheed Martin and TRW, and NASA—people who'd be likely to help build or launch or use the new telescope. The workshops resulted in a dozen or so specific ideas on what the NGST might look like and what it should do. And in September 1996, about a month before our conversation, Mather and his team got together and tried to come to a consensus.

One thing that became clear right away was that the Dressler commission, which had recommended a telescope with a four-meter mirror, had been too timid. Since building a space telescope at once bigger and cheaper than the Hubble would force NASA to be ingenious anyway, it might as well be a little more ingenious and make the telescope a *lot* bigger. The NGST, they decided, should be eight meters across, or about triple the Hubble's size.

Mather's group also agreed, as do virtually all astronomers, that the NGST should be designed to see mostly in infrared rather than visible light. All sorts of phenomena that don't give off much light in the visible-light part of the electromagnetic spectrum shine brightly in the infrared, including the birth of galaxies—the part of the Origins Program not concerned with the search for life—and the birth and development of planetary systems.

Infrared light detectors are especially sensitive to heat, which in part is why NGST will operate not a hundred miles away from Earth but a million, four times as far away as the Moon. It will probably sit at a place called L2, one of five so-called Lagrangian points where the combined gravitational effects of the Earth, Sun, and (in some cases) Moon conspire to keep an object more or less in one place. "It's like being at the top of a hill," explained Mather. "An object that's sitting there will eventually begin to roll off, but with a little effort you can keep it there. If you try to perch something on the slope of a hill, on the other hand, it takes a lot of effort to keep it in place. Getting the NGST to L2 will be like kicking a soccer ball so that it rolls to a stop right at the top of the hill. Then you need a bunch of crickets, in the form of small rocket thrusters, to keep nudging it so it stays there."

If you drew a line from the Sun to the Earth and then continued that line for another million miles, you'd be at L2. In other words, Earth lies directly between L2 and the Sun. From L2 the Sun and Earth are always in the same direction. It's therefore easy to point away from their light, which could easily fry the sensitive detectors. (Hubble's orbit is awful in comparison: The Earth always fills about a third of the sky, either blindingly bright or, when the telescope is on the night side, blocking a huge swath of sky from view. The Sun is off in another direction half the time, and there's the Moon to contend with as well. Figuring out an efficient schedule that avoids all three of these hazards is a huge logistical problem.)

The other advantage of L2 is that all the heat comes from that one direction. You can just put up a parasol, and the telescope can't help cooling down. And without the need to shield yourself from stray light, you can do away with a tube for the telescope, which makes it a lot lighter. Of course that means the mirror will be hit constantly with micrometeorites, but that turns out not to be much of a problem. You can put a lot of BB holes in a space telescope without much effect on its optical quality.

If the project keeps to its current schedule, NGST should be sailing through space, comfortably frigid behind its parasol and ready to begin operations, by 2007. Much of its work will have more to do with the origins of galaxies and such than the origins of planets. But it can look at the history of how stars and chemical elements form. It can look for the process of dust formation in very distant, very young galaxies—dust that will eventually be used to form planets.

Despite its nearly Keck-size light-gathering area and its Hubble-sharp vision, Mather admitted, it will still be hard to see planets directly with NGST. It might conceivably be done if a coronagraph was installed, a device that blots out a star and thus reveals dimmer objects near it, but it would be marginal. NGST might also be used for indirect planet searches—not just the kind Marcy and Butler do but also for astrometry, the side-to-side wobbles that Peter van de Kamp thought he'd seen in Barnard's star.

What can certainly be seen much more easily and more sharply with NGST, though, are the dust clouds around stars, a powerful test of planet formation theories. And just as important, NGST will test many of the technologies that are crucial to later, even more ambitious missions—such as exquisitely sensitive infrared detectors, lightweight building materials, thin, strong mirrors, and, most crucial, techniques for unfolding large structures in space.

The last is not, however, an absolute certainty for NGST, at least not yet. "We really have two competing imperatives," Mather said. "The astronomers want a telescope that's reliable and that can be launched soon." With a deployable telescope, the former isn't guaranteed no matter how many tests the designers do along the way, and the latter is clearly impossible. It would be simpler in many ways to build a lightweight but otherwise conventional telescope and launch it intact. But NASA wants to develop exciting new technologies that can then be transferred to industry. "The administrator is not interested," said Mather, "in having us build the same old stuff, only bigger."

Most of the experts tapped by Mather for his preliminary studies acquiesced in this prejudice. The diagrams that Mather was trotting around to display at conferences and that appeared in his reports show all sorts of ingenious ways one big mirror can be made out of many small panels, hinged and folded in on themselves like elaborate glass origami and bearing names like "stacked petal" and "spiral-fold petal."

But at least one of Mather's advisors was absolutely set against the whole idea of a deployable mirror. He figured that it was close to insane, in fact. Send the thing up in one piece, he said; why take chances with an entirely new technology? This man's opinion wouldn't ordinarily have counted for much; it was just one among the three hundred or so that Mather had solicited. Unfortunately for NASA's grand vision—though perhaps very fortunately for the astronomers who would have to use NGST—this maverick is perhaps the most talented and imaginative telescope designer working today. The MacArthur Foundation, which awarded him one of its "genius" grants in 1996, thinks so, and his colleagues generally agree. Ninety percent of the astronomers on the NGST may think the telescope will go into space folded up into a neat package. But Roger Angel will be doing his best to make sure that won't happen.

Chapter 8

THE MIRROR LAB

Peter Strittmatter, the director of the Steward Observatory at the University of Arizona, still remembers perfectly the day a young colleague named Roger Angel walked into his office and asked permission to take over some unused laboratory space.

"I asked him what for," Strittmatter told me at observatory headquarters on the university's Tucson campus early in 1997.

"Melting glass" was the answer.

"Why?"

"To make mirrors."

"What sort of mirrors?"

"Telescope mirrors. Seven-and-a-half-meter telescope mirrors."

At that point, Strittmatter recalled, "I nearly wet my pants."

This was not an unreasonable reaction. Roger Angel had no experience making telescope mirrors, aside from an abandoned attempt to build a small one as a child. He wasn't even an optical astronomer by training. His Ph.D. from Oxford was in experimental physics, and even when he moved over into astronomy as a postdoctoral student, it wasn't visible light that interested him but X rays. Angel only started dabbling in optical astronomy to get another perspective on the X-ray-emitting objects he'd been studying—superhot clouds of gas, matter squeezing down the throats of voracious black holes.

Even experienced telescope builders had never attempted to make a 7.5-meter mirror. At the time, in 1980, the largest mirror on Earth was the 200-inch-diameter (5 meter) chunk of glass at the heart of the Hale Telescope on Palomar Mountain. It had been the largest for over thirty years. Beginning in the early 1900s, telescope mirrors had grown steadily larger—thirty inches in diameter, then sixty, then one hundred. Finally, in 1949, the mammoth 200-inch-, seventeen-foot-diameter Hale saw first light in the mountains northeast of San Diego. Only one team of scientists and engineers, in the Soviet Union, even attempted to outdo the Hale. They succeeded in casting and grinding a 236-inch mirror, which they installed in an observatory in the Caucasus mountains, but it has never performed very well.

The failure to build a bigger mirror wasn't because astronomers were satisfied with the Hale's light-gathering power. Astronomers are never satisfied. In optical astronomy, light is knowledge. The feeble glow of electromagnetic radiation streaming in to Earth from the universe carries information not only about what a celestial object looks like and how hot it is, but also, thanks to spectroscopy, about how it's moving, how far away it is, and what it's made of. The more light you can catch as it falls to Earth after a journey of tens or millions or billions of light-years, the better you can understand what's going on. A telescope mirror is really nothing more than a trap that catches the light falling on its surface and funnels it into some sort of light detector. If you build a bigger trap, you'll catch more light.

That being the case, there is no such thing as big enough. No matter how big your mirror is, there's always that faint something-or-other, right on the edge of visibility, that you could understand if only you could see it just a little bit better. Unless you're talking about extremely bright objects—the Moon, the Sun, a nearby supernova—there is no such thing as too much light.

Nevertheless, astronomers let decades pass without lobbying for a telescope bigger than the Hale to be built. They hadn't by any means kicked their addiction to ever-increasing doses of light. They'd simply found a relatively inexpensive way to get it. Back at the turn of the twentieth century when Schiaparelli and Lowell were gazing at Mars, the human eye was still, as it had been since the time of Galileo, the only light detector an observer had. You looked through the eyepiece, then did your best to draw a picture of what you saw.

Shortly thereafter, astronomers discovered photography. Photographic film could also see things much too faint for the eye to detect. By opening the shutter on their cameras for long periods, astronomers could do time exposures, piling up photons upon photons, adding up light until it amounted to something perceptible.

And then, in the late 1960s, just when astronomers would ordinarily have been starting to get irrepressibly itchy about building a super Hale, they were diverted by the arrival of electronic light detectors. Photographic negatives are only about 1 percent efficient. For every photon of light they manage to trap, 99, on average, simply don't register. Electronic detectors were tens of times better; by the mid-1970s, light-gathering chips called CCDs (charge-coupled devices) were grabbing 60, 70, and even 80 percent of the light falling on them.

Suddenly, every telescope on the planet was, in effect, eighty times larger in area than it had been. A 24-inch became a Hale. The Hale became a 146-foot. CCDs didn't just make telescopes more powerful, either. Being electronic, they could dump their images directly into computers, where astronomers could process and analyze them more easily. The fact that observers no longer had to spend freezing nights sitting at a telescope's prime focus, sliding film into a camera, was just a fringe benefit. Old-fashioned astronomical photography, which a young Geoff Marcy had to learn but which Paul Butler just managed to escape, was obsolete.

CCDs kept the astronomers happy for a few years. By 1980 or so, though, they were feeling itchy once again. Why, they wondered, couldn't they have their CCDs and giant telescopes, too? There were good reasons why nobody had built the latter, and these were laid out with especial clarity in a paper written in 1979 by a friend and colleague of Roger Angel's, a theorist named Neville (Nick) Woolf.

The central difficulty of making huge mirrors, it turns out, is that you can't avoid using glass. It's cheap, it's durable, it's simple to work with, and it's extremely stable, which means it will hold any curvature you grind into its surface very well.

But glass is also heavy and retains heat. That's why a can of beer cools off much faster in a refrigerator than a bottle of beer does. As mirrors get bigger and bigger, their increasing weight forces telescope makers to design increasingly massive and expensive support structures for them. Their heat retention means that mirrors soak in the warmth of daytime air and then slowly let that warmth escape through the evening, roiling the cooler night air just as the heat rising from a car's hood roils the air above it on a summer day. And roiling air, as Woolf pointed out in his paper, makes for blurry images.

Woolf put it much more technically than that, of course. Fortunately, I was hearing all this from Angel. I had finished with Strittmatter and was heading up to the fourth floor of the observatory building, where I would spend the next couple of days visiting with both of these transplanted Englishmen. "Nick is so far ahead of everyone else," Angel told me early on, "that people don't understand what he's talking about half the time. Fortunately, I do, and it's been a very rewarding partnership."

I would come to understand what he meant. Nick Woolf is tall and slim, with a shock of unruly white hair and a high-pitched, thin voice. His eyes convey an unusual and undefinable intensity, which diminishes only slightly when he smiles—as he does often, usually without warning. If he had told me he came from another planet, I wouldn't have been all that surprised. Angel's remark about other people not understanding Nick Woolf half the time was made in reference to other astronomers. I didn't understand what he was talking about much more than half the time.

Angel's gray-white hair is unruly as well, but its rebellion involves falling down, not sticking up; it always seems to be in his face. He's rather boyish-looking (this seems to be true for male astronomers in general, as though the childhood fascination with the stars that they invariably share is still animating them) and very down-to-Earth. Angel has a gift for making the complex seem straightforward. As a hands-on experimentalist, he is the perfect foil for Woolf's visionary thinking.

One thing Woolf noted in his 1979 paper was that the problems of heat retention and air turbulence were much less at the Multiple Mirror Telescope than at most observatories. The MMT, which the University of Arizona shares with the Smithsonian Institution, sits on top of Mount Hopkins, forty miles south of Tucson, on the highway to the Nogales and the Mexican border. As its name implies, it has, or had at the time, six small mirrors where other telescopes have one large one. The six were surplus spy-satellite mirrors donated by the Air Force, and unlike just about every other mirror at the time, the mirrors weren't solid. They were a sort of sandwich made from two round plates of glass attached to a waffle-patterned glass structure. The air spaces inside the waffling helped the mirrors cool off to nighttime temperature very quickly. Why, wondered Angel, couldn't they build much larger mirrors the same way?

So he began to play with glass, first in a makeshift oven at home and then, with Strittmatter's encouragement, at the university. Finally, in 1985, the astronomy department, under Angel's incessant urging, persuaded the university to build an entire laboratory devoted to mirror-making. It was installed snugly under the north stands of the university's immense concrete football stadium, a five-minute walk from the observatory.

By now Angel had built mirrors as large as 3.5 meters across, and he and Woolf had decided that their ultimate goal would be an 8-meter monolith. This figure had nothing to do with either science or engineering but with the fact that most two-lane highways are only fifty feet wide. A mirror is useless if you can't get it to the telescope; anything bigger than 8 meters across, loaded onto a flatbed truck, would never make it out of Tucson. So 8 meters would be their limit.

Before they were prepared to make a mirror this gigantic, Angel and Woolf wanted to take one last intermediate step. They wanted to trump the Hale. They would do it by casting a mirror 256 inches, or 6.5 meters, across, a little more than 4 feet larger than the one at Palomar. If it worked, this record-breaking slab of glass would be installed in the MMT to replace the six small mirrors that were in a sense its inspiration. As we sat in Angel's office, I asked him whether they'd pulled it off. "Would you like to see it?" he asked modestly. "It's just about finished."

The 6.5-meter was sitting at that moment in the Mirror Lab's cavernous polishing room, a chamber easily eighty feet high and perhaps one hundred feet square. You could stack three or four two-story houses inside. Large as the mirror was, it looked tiny inside this huge space—a hockey puck of glass, its top surface gently concave, as though a giant had pushed it in slightly with his thumb.

After the Hale Telescope builders had put the final touches on the 200-inch mirror, they breathed a figurative, collective sigh of relief. When Angel was about halfway done with his 256-inch mirror, he turned around and started building another. As a result, the polishing room held not one 6.5-meter slab of glass but two. The second sat over in the far corner, a hockey puck identical to the first but without a precise light-gathering curve burnished into its surface.

When it was completed a year or so later, this second 6.5-meter mirror would become the heart of the Magellan Telescope, which is being built in the Chilean Andes by the Carnegie Observatories. The University of Arizona will get a piece of the Magellan as well. For now, the mirror-in-progress sat on a large turntable with a heavy white girder laid across the top of the mirror, perfectly centered. As the turntable spun, a diamond-tipped grinding bit would move in and out along this axis, shaving excess glass from the blank.

Up close, it was obvious that these yard-thick slabs of glass weren't solid. Both mirrors had a top layer, concavely curved and about an inch thick. Below that, easily visible through the not-yet-silvered glass, was a distinct honeycomb pattern. It marked a network of hexagonal chambers about eight inches across, separated by walls an inch or so thick that stretched from just under the top layer of glass nearly all the way to the back of the mirror. These chambers—about a thousand of them, filling the entire volume of the slab—were full of air. The mirror is mostly empty space.

As a result, it is not only much lighter than a conventional mirror—ten tons as opposed to about fifty if it were solid glass—but it will also have no problem with heat retention. A hole has been drilled into each chamber through the back of the mirror. When the telescope is in operation, dome-temperature air will be pumped continuously into each of the one thousand hexagonal cells through the holes. The day's heat won't have to leak slowly from the core of the mirror to its surface and out into the surrounding air; it will be sucked away. The mirror will be centrally air-conditioned.

Unlike the Air Force's mirrors, though, whose design Angel ultimately rejected as too unwieldy for large mirrors, these had been built with an entirely new technique, a MacArthur-worthy flight of Angel's genius. As Angel led me through a doorway into the room where his mirrors are actually cast—and where his ultimate creation, an 8.4-meter, 27-foot monster mirror was even now cooking in the oven—he handed me what looked like a chunk of Styrofoam. It was shaped like a hexagonal cylinder—a slice of a number two pencil—8 inches across and about 18 inches long.

In fact, I was holding a brick made of ceramic fibers, the same terrifically temperature-resistant material that NASA uses to make the tiles that heat-proof the space shuttle. When Mirror Lab technicians are ready to cast a new mirror, they fill a mold with these bricks, standing them on end atop ceramic disks and separating them by about an inch. Then they pile chunks of glass on top of the bricks—41,000 pounds' worth for the 8.4-meter mirror—and melt them at 1180 degrees Fahrenheit; the glass flows down and around and under the bricks, forming a pool of glass beneath them, filling all the spaces between them, and overflowing to form a shallow pool of glass on top of them.

After the glass has cooled and hardened, the technicians take high-pressure jets of water and literally pulverize the ceramic—first the disks underneath and then the bricks themselves. Brick dust and water flow out of what are now three-inch holes in the back of the mirror. What is left are two plates of glass with a honeycomb of glass stretching between them—the structure I had seen in the nearly finished mirrors next door.

Figuring out how to carve out most of a mirror's insides was Angel's first innovation. The second was to melt it inside a rotating oven. The effect is identical to what happens when you stir a pot of soup vigorously. The liquefied glass sloshes out to the edges of the container, leaving a well in the middle. The surface assumes the geometric curve known as a parabola, the ideal shape for a light-focusing mirror. Because the glass keeps rotating as it cools, this parabola is frozen permanently on the surface of the glass. Spin-casting has at a stroke eliminated an entire step in the mirror-grinding process. Angel estimates that it has saved him two years of work on the 8.4-meter mirror and thirty tons of glass that now would not have to be carved away.

At 6.8 revolutions per minute, the rotating oven will make the high edge of this particular mirror about three feet thick, the middle about half that. This is deeper than the curvature of most mirrors, which is a good thing. A deeper curve means the mirror will focus light relatively close to its surface—just a little over eight meters away. The ratio of focal distance to mirror width is called the focal ratio, and the smaller it is, the better. When the focal ratio is short, light comes to a tighter focus; it's more concentrated, and the image is brighter. It is much harder to grind a deep curve than it is a shallow one. Before Angel perfected his spin-casting technique, the best anyone had done with a telescope mirror was F 1.8—a focal ratio nearly twice the mirror's width. The mirror cooking in Angel's oven would have a focal ratio of F 1.14.

As Angel was explaining all this, a thought struck me. He'd just told me a little while earlier that he and Nick Woolf had settled on eight meters, no more, as the maximum size of a mirror. The oven turning on its axis across the room held a mirror nearly half a meter bigger than that. What was the story?

"Ah, well," he said, looking a little sheepish, "there's been something of a pissing contest going on." Angel is not the only mirror maker in the world, not even the only one turning out large mirrors. The European Southern Observatory, run jointly by several European governments, is building a facility in Chile called the Very Large Telescope that will have four 8-meter telescopes working in concert. The Japanese government is building an 8-meter telescope, the Subaru, on Mauna Kea, within shouting distance of the Keck. The U.S. government, in partnership with six other countries, is building two 8-meter telescopes, known as the Gemini Project, one on Mauna Kea and one in Chile.

"Of course," said Angel, "everyone wanted to have the biggest mirror." So while it was generally agreed that eight meters would be the optimum size, the Europeans decided that their 8-meter mirror would really be 8.2 meters across. This convinced the Japanese that theirs should be 8.3. In retaliation, Angel upped his to 8.4.

Angel's will be the biggest single-piece mirror on Earth, then, but only by about four inches. He still thinks it will perform significantly better, though, due to its short focal length, and be far cheaper to operate than the competition. All of the other big-mirror shops, including Corning Glass in New York State and Schott in Germany, have taken a very different approach to breaking the Hale barrier. The VLT, the Subaru, the Gemini—and, in a slightly different sense, the Kecks—have all opted to cut down on weight and heat retention not by making their mirrors thick and hollow but by making them thin. Where Angel's latest mirror is three feet thick at its edge, the VLT mirrors, nearly as wide, are only about eight inches thick; the other big mirrors will be comparably thin.

As a result, these wide, thin mirrors aren't even close to being stiff. Left unsupported, they'd flop like pancakes. The only way to hold their curvature is to support them at dozens or even hundreds of points along their back surfaces, with individually adjustable mounts that can force the mirrors back into shape when, inevitably, they sag and warp under their own weight or bow in the wind like spinnakers. (If the Keck mirrors were unsupported, they'd be in thirty-six pieces each, lying on the ground; the principle, though, is the same.) "Our mirrors are a lot more complicated to build," admitted Angel, "but once they're finished, they're much easier to control." Perpetual computer-controlled fine-tuning is the price the Kecks and the VLT and the rest will have to pay forever for their thinness.

In space, on the other hand, a mirror doesn't have to contend with wind or gravity. Thinness isn't such a problem, and if taken to extremes, it can be a real advantage. A very thin mirror—thinner than the VLT and Gemini mirrors by far—will necessarily be extremely light and thus ideal for launching into space. NASA's NGST will most likely incorporate an unprecedentedly thin mirror.

This is a straightforward enough conclusion, arrived at independently by a number of scientists. Roger Angel did, too, but characteristically, he did so in a roundabout fashion. The giant mirror cooking in the Mirror Lab's rotating oven during my visit was one of a pair destined for a radically new sort of telescope under construction on Mount Graham, Arizona. When it's complete in 1999, the Large Binocular Telescope will carry twin 8.4-meter mirrors riding on a single mount. They'll have a total light-collecting area equivalent to that of a single telescope 12 meters in diameter.

Like many telescopes now being designed—NGST included—the LBT will be sensitive mainly to infrared light, the part of the electromagnetic region where astronomers believe the universe has the most secrets left to reveal. Because it will be working from the ground, though, the LBT will constantly be fighting the background infrared radiation that leaks from any object—trees, telescope, buildings, people—whose temperature is much above absolute zero. Accordingly, Angel and Woolf wanted to minimize any possible sources of unnecessary infrared—extra pieces of hardware, for example.

They hit on a radical idea. No respectable ground-based telescope is built anymore without an adaptive-optics system. This is a scheme that attempts to compensate for some of the inherent shimmering of the atmosphere—the thing that makes the stars twinkle—by bouncing starlight off an extra mirror on the way to the detector. The extra mirror is flexible, rubbery, and flexed in just such a way as to counteract the blurring effect of the roiling air above. It roils in lockstep with, but in the opposite direction of, the atmosphere.

Adaptive optics, perfected by the Air Force years ago and declassified in the early 1990s, works fine, although it can't eliminate all of the atmospheric twinkle. But the extra mirror it would force them to install had Angel and Woolf wringing their hands over yet another source of excess heat and infrared. "We were thinking about it," said Angel, "and we asked ourselves, 'Why not make the atmospheric corrections directly on the secondary mirror?'"

Angel was referring to a piece of hardware that is an integral part of most large astronomical telescopes. The secondary mirror takes light collected and focused by the big primary mirror—the sort Angel casts in his mirror lab—and bounces it back down into the observatory to be recorded and analyzed. It's the piece of hardware that lets a reflecting telescope be only half as long as the equivalent-size refracting telescope. Each of the LBT's two 8.4-meter primary mirrors will have a secondary mirror no matter what. All Woolf and Angel had to do was to figure out how to make mirrors with the high optical quality necessary in a secondary, but make them flexible enough for adaptive optics. Instead of two bounces after hitting the primary mirror, light would take just one.

Nobody had done such a thing before, however, and so Angel turned to a Mirror Lab staff physicist named Buddy Martin, a low-key, completely self-effacing young man who had solved some of the lab's messier problems over the years. "Roger had the basic idea," Martin told me. "I just worked out the details." You can make a piece of glass very thin, Martin explained. That's easy. But you can't grind the necessary curvature into it unless you can mount it on something substantial first. Moreover, the mounting has to support the glass equally at every point; otherwise, the slightest pressure from a grinding tool will crack it.

What Martin and Angel decided to do was mount the glass first and make it thin afterward. They took two disks of glass, each four inches thick and a half meter across, and ground spherical curves into their surfaces—one convex, the other concave, so they'd fit together perfectly. Then Martin and his engineers glued the two pieces of glass together, the convex one on top, using pine pitch. Pitch is the ideal glue for situations like this because it's sticky, spreads out to make a thin, even layer between the two pieces, and can be softened easily by heating it if you want to pull them apart again.

Once the two disks were glued together, Martin's team began grinding away at the top one. They kept grinding, removing a millimeter of glass at a time, until all that was left was a sheet 2 millimeters thick, with matching perfectly spherical curves in both its top and bottom surfaces. It was basically a contact lens 19.5 inches across and about as thick as a nickel. "If we were making an actual secondary mirror," Martin explained, "we would then have gone on and ground a hyperbolic curve into one side. It would have been pretty easy."

This was just a test of the concept, though, so they went directly to the following step. They warmed up the glass until the pitch softened, tipped the whole thing, and the thin glass shell just slid out. "It was a pretty straightforward thing to do at a half meter," Martin said. "It should be real interesting to handle when we get up to six meters." As long as they support the glass evenly across its whole surface at all times, though, he expects the technique to work.

Finally, to demonstrate how they would manipulate its shape once it was installed in the telescope, Angel and Martin mounted the glass on actuators—essentially electrically controlled screws—embedded in a disk of carbon fiber and epoxy composite. Each of the dozen actuators can move the mirror in or out almost instantly by as little as ten billionths of a meter, a degree of control that makes them adequate for adaptive optics.

If they could build a large, super-thin, optically-near-perfect secondary mirror for the LBT, Angel realized, they could use the same technique to build a primary mirror. Where actuators would be used to change the shape of an adaptive-optics secondary from moment to moment, they could be used equally well to keep a primary in precisely the same shape at all times.

He wouldn't want to do it for a ground-based instrument, since even a moderately thin mirror goes against his basic telescope-making philosophy. But Angel knew that the NGST's success depended on a very light mirror. He had what might be the answer to John Mather's dream. "The Hubble telescope, with its two-and-a-half-meter mirror, weighs eleven tons," Angel said. "A six-meter NGST with one of our mirrors would weigh only two tons." Strittmatter, whose support and administrative skills Angel credits for much of the lab's success, invited Dan Goldin to see the thin-mirror prototype. He and Angel set it up so that Goldin could operate it with a joystick, moving it into and out of shape. "He seemed very impressed," Strittmatter said.

This is not to say that either Goldin or Mather accepts Angel's proposition that the NGST should go up whole rather than unfold in orbit. "If someone says, 'I can do it,' I'll say, 'Prove it,' " Mather told me. "If it's feasible, terrific, but the people who are advocating it haven't done very much to demonstrate that." Besides, he said, "the whole idea bothers me because it's the wave of the past, not the future. I'd really prefer to do something more clever."

That, Angel believes, is the problem. He's convinced that this prejudice in favor of cleverness has distorted the way people at NASA think. "Deployment technology is all very exciting," he said, "but as an astronomer I want NGST *not* to be a guinea pig for new technology. I want it to be a real instrument. I don't want us to design something that will second-guess what we'll want to be able to do twenty years from now. Let's build the best telescope we can launch right now." In response to Mather's claim that he hasn't done much to demonstrate his concept's advantages, Angel had completed, just before my visit, a thick position paper that he said he was about to forward to Goddard.

Even a visionary like Angel admits that a one-piece eight-meter mirror is too large to send into space. He also acknowledges that the United States has no rocket big enough to hold even a six-meter mirror, his own candidate size for the best space telescope we can build right now. But the European Space Agency has its own rocket, the Ariane, which is already five meters in diameter. "You could easily modify Ariane or even the Russian Proton rocket to hold a six-meter telescope," he said. "It turns out that the cost of doing that would be only about $100 million."

Angel also admits that the idea of going out of the country shopping for rockets is not something NASA will be comfortable with. His thin mirrors may well be used in NGST, just not in the way he would prefer. Whatever happens, he said, "I have to say that this has been a lot of fun, figuring all this out. When you see something exciting like this, you can't easily let it go. I've got to settle down, though." He had the world's largest single-piece telescope mirror in the oven, after all, and another just as big to cast right afterward.

The planet-hunters aren't so much interested in clever tricks—Angel's or NASA's or anyone else's. They simply want as much light as they can get to explore as many stars as quickly as possible. They need planets by the hundreds so that they can begin to understand the natural history of planets at a deep level. And they need ones that are smaller than anyone has found yet. 51 Peg Bs and 70 Vir Bs and the like may eventually put a number F_n in the Drake equation, the factor that says how many stars have planets. But they won't be able to say much about F_e, the number of planets in the average solar system that are hospitable to life. To do that, they won't necessarily have to find Earths, at least not right away, but until they can plumb the cosmos for objects the size of Jupiter and Saturn and Uranus and Neptune, and find them orbiting at about where they do in our own solar system, planetary scientists won't be able to say with much confidence how many stars are likely to have Earths circling them.

Even the most powerful telescopes in existence, though, won't be able to detect Earths by radial-velocity measurements. The reason is that Earth's effect on the Sun is tiny. Giant Jupiter makes the Sun wobble at a speed of twelve meters, or thirty-nine feet, per second. Tiny Earth makes it wobble by ten centimeters, or about four inches, per second. No radial-velocity search will ever detect wobbles that feeble. That's partly due to the fact that the accuracy Butler and Marcy have achieved is at the very nail-biting edge of what's possible. But if they could improve their precision significantly, it still wouldn't do any good. Even stable stars like the Sun pulsate a tiny bit— enough to hide any signal an Earth could impose. "For the next two or three years," says Butler, "radial velocities like we do are still going to be the game. After that, we're going to get our butts kicked."

Chapter 9

BEYOND
THE TELESCOPE

When Daniel Goldin presented NASA's Origins Program to the American Astronomical Society in San Antonio in January 1996, the Next Generation Space Telescope was only one of the futuristic instruments he described.

In many ways it was the least interesting. The chief advantage of lofting a telescope into space is that there's no atmosphere to blur its images. A space-based telescope can take preternaturally sharp pictures of heavenly objects, revealing their fine details. The ability to keep such details from blurring is called resolution, and it is the key to understanding all sorts of astronomical phenomena. The strongest evidence for the existence of black holes comes from super-sharp Hubble pictures of gas clouds swirling at the very cores of faraway galaxies. Proof that galaxy birth is a tumultuous, almost chaotic process comes from Hubble images that vividly reveal the tortured shapes of nascent star clusters.

It is high resolution that will ultimately give astronomers the additional information they need to plumb the secrets of the Drake equation as well. As Marcy and Butler's radial velocity measurements reach their sensitivity limit, the competing method for finding planets will become ascendant. Astrometry—the search for side-to-side rather than forward-and-back wobbles, the technique Peter van de Kamp thought he'd mastered—is ultimately much more sensitive than radial velocity searches. It can, in principle, detect the existence of planets as small as Earth.

That's far beyond the capabilities of the Hubble, though, and even beyond the capabilities of the NGST. Lofting a telescope above the Earth's atmosphere removes one source of blur, but a telescope's resolution is also limited by its size. All other things being equal, a small telescope cannot focus light as sharply as a large one. An eight-meter NGST simply does not have the resolving power to pick out the effect of Earth-size planets. A one-hundred-meter NGST might, but such a device is beyond rational contemplation.

Yet astronomers are confident they can achieve the necessary resolution, not by building impossibly large telescopes, but by resorting to a sort of optical trick. It's a process called interferometry; indeed, the two instruments that Daniel Goldin promised the Astronomical Society, in addition to the NGST, were labeled the Space Interferometry Mission, or SIM, and the Planet Finder Interferometer.

In its simplest possible description, interferometry is the combining of light from two small, widely separated telescopes to achieve the resolution of a single huge one. The principles underlying that straightforward statement are rather more complicated to grasp, and the execution—the workings of an actual interferometer—more complicated still.

Mike Shao, an astrophysicist at the Jet Propulsion Laboratory, is the scientist in charge of NASA's interferometry program. He works on Palomar Mountain, the home of the two-hundred-inch Hale Telescope. The dome of the two-hundred-inch still dominates the mountaintop, a sort of national parkish place covered, in jarring contrast to the moonscape of Mauna Kea, with pine trees, meadows, and plenty of wildlife (the mountain lions generally stay outside the facility's chain-link fence).

Sitting in the shadow of this monument to astronomy's glorious past are three white utility sheds, one large and two small, connected by white metal pipes that are suspended about four feet off the ground. This unprepossessing setup, which looked like nothing so much as an air-conditioning plant, is in fact a monument to astronomy's hopeful future. It's the so-called Palomar Testbed Interferometer, a ground-based working prototype of the device Dan Goldin wants to launch into space by the year 2003. These sheds and their contents, funded by NASA, are some of the Origins Program's first hardware.

None of this would become clear to me until late in the afternoon when Mark Colavita showed up. Colavita is Shao's right-hand man, a brilliant scientist in his own right and an important contributor to the interferometer's success. Colavita is the perfect model of a nerdy scientist: He's extremely thin, has a large head of curly black hair, and tends to speak in long, technical sentences in a voice that rarely rises above a murmur. Like Shao, though, he's happy to explain things again in case you still don't get it the fifth time around.

As he began setting up the interferometer for the night's operations, Colavita explained the theory behind interferometry in a little more detail. As discussed earlier, light travels in waves, little peaks and troughs, of electromagnetic energy, analogous to the waves of an ocean. The distance from one peak to the next, or one trough to the next, is called the wavelength. Visible light has wavelengths ranging from about four hundred nanometers — that is, four hundred billionths of a meter — for the most violet of violets, and up to eight hundred nanometers for the reddest reds. Longer wavelengths than that take you into the infrared light, then microwaves, and finally into radio waves, whose peaks can be feet or even yards apart. Wavelengths shorter than four hundred nanometers correspond to ultraviolet light, then X rays, then gamma rays.

Now, what happens if you take two distinct beams of light and combine them? It depends on how their waves overlap. If the waves are aligned perfectly so that peak overlaps with peak, you get, in essence, a double peak — a light twice as bright as each beam by itself. If the waves are perfectly out of alignment, on the other hand, the peaks from one beam will overlap the troughs of the other. Peaks cancel troughs; the result is darkness.

None of this is noticeable in the ordinary world because the light we see bounces off so many surfaces and is refracted through so many air molecules that nothing overlaps anything with enough precision to make a difference. In a laboratory, though, you can manipulate light beams to produce a pattern on a detector that alternates strips of light and dark—places where the beams are alternately canceling and reinforcing each other. (The two phenomena are formally known as destructive and constructive interference; the strips are called "fringes.")

If you adjust the overlap by fractions of a wavelength—distances on the order of a hundred-thousandth of an inch—you can make the center of the pattern especially bright through constructive interference, or quite dark through destructive interference.

It's this effect—minuscule shifts in the positions of light beams that result in huge brightness variations—that makes interferometry so attractive for astronomy and, in particular, for astrometry, the measurement of tiny shifts in a star's position. If you take two telescopes separated by a few dozen feet, aim them both at the same star, and combine the resulting pair of light beams, you will create an interference pattern. If the star changes position by even an imperceptibly small angle—pulled by a planet's gravity, for example—the interference pattern will shift dramatically.

Planets aren't the only thing that can make stars move or seem to move. For example, if you check a star's position in, say, January and then again in July, the Earth will have moved to the other side of the Sun. You'll be looking at the star from a different angle, and it will appear to have shifted its position relative to other, more distant stars. (You can get the same effect by holding a finger upright a foot in front of your face and looking at it first with one eye, then the other. The finger seems to move; what's really changing is just the angle of view.)

The angle by which a star appears to move over a six-month period is known as its parallax, and the degree of parallax tells you exactly how far the star is from Earth. A closer star will seem to move farther, just as a closer finger jumps more violently as you switch eyes.

The best astronomers can do with current technology is measure parallax to about one thousand light-years; any farther than that and the changes in angle are too small to measure with conventional telescopes. A space-based interferometer will be able to measure angles that are a hundred times smaller and therefore will be able to measure parallax, and thus stellar distances, all the way across the galaxy.

More to the point for the Origins Program, you can also see with high precision whether a given star is wobbling from side to side. There's an enormous gap, however, between the principle and a useful device. The first interferometers were built more than a century ago, but they didn't have anywhere near the accuracy needed to measure planetary wobbles. In intervening years, astronomers periodically returned to the problem, but never with much success (except, that is, with radio telescopes; radio waves are much longer than light waves—feet rather than millionths of an inch—and the precision required to combine them is correspondingly less).

Mike Shao is one of a handful of astronomers who are finally managing to make optical interferometry work. His persistence and care in doing so has earned him a place on Paul Butler's good-guy list despite, or maybe thanks to, the fact that Shao is the one who will kick Marcy and Butler's collective butt. "He's kind of like us," Butler told me. "He's been working for ten years or more on this project. He has run into levels of possible errors that people have never conceived of before."

As Colavita started preparing the Palomar interferometer for the night's observing, the dome on the two-hundred-inch slowly and majestically opened to reveal the huge telescope inside. Opening the interferometer wasn't nearly such an impressive operation. It mostly involved sweat. He hauled one of the sheds aside—the entire thing, except the floor, was mounted on wheels—and then the other, to reveal a small telescope mounted in each. These were the edges of the interferometer, the individual telescopes whose light would be routed down the white metal pipes and into the main building to be optically combined.

At the same time, a laser beam would be shooting out from the main building in the opposite direction, backtracking over the light path to make sure the mirrors are all aimed properly—that light will, in fact, pass through the system. It was already shining in preparation for the night's work. Standing in one of the small sheds, Colavita put a white card near the pipe opening to make sure it was aimed correctly. A scintillating red dot appeared—a little off center. It would have to be adjusted. I leaned over and looked down the pipe toward the source of the laser beam. There, glowing hotly from the building at the other end, was a brilliant spot of red light. Too late, I realized that this might be a really stupid thing to do. I jerked my head back violently. "Don't worry," Colavita said, laughing. "You'd have to stare at it for a while to do any damage."

Inside the main building, Colavita showed me the heart of the interferometer, an impossibly complex-looking collection of mirrors and detectors and actuators—enough equipment to give nightmares to someone like me, who barely made it through high school physics. I asked Colavita how many bounces the light actually has to make before its journey is over. "Um . . . I don't know," he said. "I could probably figure it out. Twenty?" I observed that this looked a lot more complicated than any telescope I'd ever seen. "It sure is," he said proudly. "People who are next door working on the two-hundred-inch come over here to take a look around. They can't quite believe it."

Yet all of this complexity is necessary if you're going to keep two light beams aligned to within a millionth of an inch, the margin of measurement safety that Shao and Colavita need if they are confidently to measure changes of a hundred-thousandth of an inch in either direction and to recheck the alignment every ten milliseconds. Moreover, this interferometer is designed to measure the positions of two stars at a time; you can't tell if a star is wobbling unless you compare it with another star. So the Palomar Testbed is actually two interferometers in one, sharing light paths but looking independently at different objects. That means the astronomers have crammed two sets of everything—mirrors, lasers, and so on—into their apparatus.

(One obvious question: Since you're measuring one star in relation to another, how do you know it's your target star that's wobbling and not your reference star? The answer is that if you find a wobble, you check again against a different reference star.)

The Testbed had been operating for about a year and a half when I visited, but this more complex dual-star observing system had been up and running for only about a month. It would be many weeks more before Shao and

Colavita could add their next upgrade: a third telescope with its own shed and pipes and mirrors and delay lines and lasers. A pile of construction materials—light pipes, prefab building panels—lay in the driveway, ready for a crew to assemble.

By installing a third arm on the interferometer, at right angles to the other two, the astronomers would be able to measure not just one direction of wobble, the way Marcy and Butler do, but two—north-south as well as east-west. They would be able to deduce not just the period and the eccentricity of whatever planet-induced motion might be present but also the shape of the orbit as it appears in the sky. If a planet is orbiting face-on, they'll know it because the shape will be fully circular (or as fully circular as the orbital eccentricity allows). If it's nearly edge-on, that will be obvious, too. Knowing at what angle they're looking will let the astronomers know how much of the wobble they're missing, which will let them calculate the planet's mass unambiguously.

In the control room itself, Colavita introduced me to Xiaopei Pan, a Caltech astronomer. Pan's specialty is stars, and he's been using the interferometer to study them. The fringes created by a star's interference patterns carry information about a star's position and about its size.

Mike Shao then arrived and proceeded to tell me about how he got into this maddeningly complicated branch of astronomy. "In modern history," he explained, pausing frequently to push his glasses back up his nose, "we were the first ones in the United States to get serious about interferometry. I started building the Mark 1, my first attempt, in 1979 when I was in grad school at M.I.T." He went on to build another, more sophisticated device and then, in the mid-1980s, a third. Until then, interferometers had been so complicated that only the people who built them could use them. It took hours just to get them aligned. The Mark 3 was the first one to incorporate computers powerful enough to do some of the alignment work.

Seated at the keyboard that controls the Palomar Testbed Interferometer, Xiaopei Pan was unwittingly giving emphasis to Shao's words. The PTI is more powerful and more computer-controlled than any of the earlier models. It goes out on its own, detects the brightest fringes—in seconds, usually—and then sits there, tracking the star, until it's told to stop. "Basically,"

Shao explained, "Xiaopei punches in his list of target stars at the beginning of the night, and the interferometer just runs through them."

Sophisticated as it is, the Palomar Testbed Interferometer is, as its name implies, just another experimental device, a step along the road to even more powerful machines. Once the PTI begins looking at the sky in earnest, it should be able to equal but not surpass Butler and Marcy's performance with the Hamilton Spectrograph at Lick Observatory. It should detect planets like 51 Peg B and 47 UMa B and 70 Vir B.

It won't find Jupiters, though. At a distance of thirty-five light-years away, Jupiter makes the Sun move by a total of about one milliarcsecond—one two-millionth the width of the Moon—over an eleven-year orbit. Uranus makes it move about 160 microarcseconds, or more than ten times less than that. The PTI could detect Jupiters, though not Uranuses, but there's a catch even to finding Jupiters. In order to detect a planet, you need a nearby reference star. But the PTI's mirrors are too small to see any but the brightest stars, and you can't count on finding a bright reference star close enough to the target star you're looking at. Ideally, you want to build your interferometer with the biggest telescopes you can.

As it happens, just such a device is well along in its design. If Mike Shao had stuck with his original naming scheme, this new device would be called the Mark 5. It's actually called the Keck Observatory.

"Interferometry is actually one of the huge reasons they built Keck II," Butler would tell me a few months later as we walked down the dimly lit corridor that connects the two domes. Beneath our feet was a second corridor, a tunnel where the light from the two telescopes will someday be combined. Interferometry is also one reason that NASA kicked in $30 million for its construction and now claims thirty Keck nights a year in dividends. Eventually, a series of much smaller outrigger telescopes will be hooked in to the two big ones, to give the Keck interferometer the same sort of two-dimensional sky coverage that a three-arm Palomar interferometer will have.

Shao and Colavita have already been to Hawaii several times, scouting the location and trying to anticipate any problems they might have. "It's already clear," Shao said, "that working with the ten-meter telescopes is going to be a challenge. Last year we went out there with accelerometers to see what happens when it gets windy. Those big ten-meter telescopes vibrate too much. We're going to have to deal with that."

It will be worth it. The Kecks are so exquisitely sensitive to light that finding suitable reference stars won't be difficult at all. Dim stars that can't even be seen with the PTI's puny telescopes will be dazzling through the Kecks. Beyond that, the Keck telescopes are now being outfitted with their own adaptive optics systems, less sophisticated but otherwise similar to the one Angel is designing for the Large Binocular Telescope. It's a lot easier to measure a star's position when it's a sharp pinpoint than when it's a blur. "Adaptive optics alone," said Shao, "should buy us an extra four-hundred-fold improvement in sensitivity." With that sort of performance, while the PTI will be able to detect 51 Pegs, the Kecks should, in Ed Weiler's words, "find Jupiters like crazy and maybe Saturns and Uranuses."

So, for that matter, should any large telescopes that can be turned into an interferometer. The Very Large Telescope with four 8-meter telescopes now under construction in Chile was, like the Kecks, designed to work that way. So is the LBT—although, once again, Angel argues persuasively that his creation has big advantages over the competition. Not only will the LBT have flexible secondary mirrors that will eliminate a major source of heat and infrared contamination, but it will also have much less hardware of all kinds.

That's because, unlike all the other large-telescope interferometers, the LBT has both its mirrors installed in a single mount; they'll aim and point as a unit, like a pair of binoculars. With independently mounted telescopes, a beam of light has to travel at least a few feet farther to reach one of the telescopes, especially when the object being looked at is near the horizon. If these beams are going to be combined to make a useful interference pattern, this inequality has to be equalized first—and equalized over and over as the Earth turns and the discrepancy constantly changes.

A good fraction of the nightmarish tangle of optical equipment in the Palomar interferometer is there just to take care of this problem, and the Keck interferometer and the VLT interferometer will have to address it as well.

Not LBT. Dispensing with an entire level of extreme complexity—a whole set of moving mirrors, feedback loops, computer controls, and so on—should make the LBT significantly more reliable than its rivals. Eliminating hardware also means eliminating extra sources of heat and, thus, of infrared radiation that can swamp the faint signals streaming in from alien solar systems.

By 2003, however, presuming that the dust problem isn't too severe, the Space Interferometry Mission should make all ground-based interferometers more or less obsolete. When Ed Weiler told me about NASA's plans for Origins, I got the impression that SIM, the Space Interferometry Mission, was something that had only been discussed in a vague way. When I asked Shao when SIM might get started, he looked utterly baffled. "We've been working on it actively for five years," he said. Surely, though, this was all still in the planning stage. There was no actual hardware. "On the contrary," he said, "we have a working model already."

The model, as I saw when I got down off the mountain and over to the Jet Propulsion Lab in Pasadena, is a three-armed gridwork structure that hangs from the ceiling of a wide two-story room. It's not a scale model; each arm is about twenty feet long, which means the device—officially known as the "micro-precision interferometer testbed"—is just as big as the interferometer that will fly in space.

Because the space interferometer will be operating in free fall, Shao and his team face different sorts of problems than they have at Palomar or will have at Mauna Kea. On the ground, an interferometer can be jarred by passing trucks, distant earthquakes, and strong winds. In space, none of these is an issue, but space creates other problems. The entire interferometer has to be as light in weight as possible to save on launching costs, but that means its structure will flex at the slightest opportunity. The motors that move the mirrors back and forth to keep fringes at their brightest, for example, send almost imperceptible shivers through the device. Even a very slightly stronger pull of Earth's gravity on one side of an orbiting interferometer than on the other can set up vibrations.

The other chief difficulty in space is one that Shao has imposed on himself. On the ground, even adaptive optics can do only so much to counteract the atmosphere's inherent blurriness. In space, there's nothing to counteract.

Stars are truly pinpoints of light whose positions can in theory be measured to absolute precision. In practice, however, astronomers are limited by the precision of their equipment. An interferometer's accuracy depends on how precisely its operators can measure the lengths of the light paths. On Palomar Mountain, Shao and Colavita have impressed their colleagues by making those measurements to within .000001 inch. In space they want to take advantage of the clarity of the seeing by measuring their mirror motions to within .000000001 inch. That means they should be able to measure wobbles of only a few microarcseconds of angle in nearby stars—a sensitivity that will finally allow astronomers to detect the wobbles caused by distant Earths.

In order to test their measurement devices as accurately as possible, explained Gregory Neat, a tall young scientist on Shao and Colavita's JPL team, the scientists first need to simulate the space environment in which SIM will actually live. The prototype is therefore suspended from bearings that literally float on pressurized air. That, said Neat, keeps out the gross vibrations caused by local thumping, which are irrelevant to the kinds of shaking they'll encounter in space. The scientists were still having trouble with air currents pushed around by the building's ventilating system, so they were building a box around the whole thing to keep the air inside perfectly still.

What's left after terrestrial influences have been rooted out will be the sorts of minute but irritating vibrations that happen in space. These can't be eliminated since they're an inherent part of the interferometer, but they can be counteracted. To do so, Shao's group has fitted out the test interferometer with voice coils, the same sort of electromagnetic "shakers" that set air vibrating in stereo loudspeakers. The coils vibrate the structure deliberately, but in the opposite sense of the unwanted vibrations. Where the interferometer zigs, the motors force it to zag, at the same time and with the same intensity. The result is no motion at all. It's destructive interference—peaks meeting troughs, troughs meeting peaks so that everything cancels—except that instead of electromagnetic waves, these are mechanical waves in the hardware.

So far the scientists and engineers working on the JPL test interferometer haven't run into any difficulties they don't believe they can handle, unless you count the huge, unwieldy administrative infrastructure that dominates the U.S. space program. The kinds of measurements Shao and Colavita propose to make with SIM sound absurd in their precision, but no less a skeptic than Paul Butler insists that "if it's physically possible, Mike Shao is going to do it."

If SIM does make it and lives up to the performance standards Shao and Colavita have set for it, another factor in the Drake equation will be quantifiable. Marcy and Butler and their planet-hunting colleagues should already have nailed down F_p, the fraction of stars harboring planets. With its ability to detect Earths, SIM will be able to do the same for F_e, the fraction of planetary systems with worlds suitable—in an average sense, at least—for life.

But to take the next step—to solve F_l, the fraction of suitable planets on which life has actually taken hold—you can't rely on indirect detection methods, on radial velocities or astrometric wobbles. You have to look directly at the planets themselves. You have to take pictures of them, analyze their light, and search their atmospheres for the gaseous waste products that living organisms give off.

That job, should NASA manage to keep Congress and the American people interested through the launching first of SIM and then of the NGST, will fall to the third orbiting observatory that Daniel Goldin described to the Astronomical Society meeting in San Antonio, the instrument he called "Planet Finder." It, too, is an interferometer. Where Mike Shao's SIM is designed to measure the back-and-forth wobbles in stars, the Planet Finder will actually make images of otherworldly planets.

Once again, as they have with big telescope mirrors, ground-based interferometers, and thin mirrors for the NGST, Angel and Woolf will have a major influence on the concept and design of the Planet Finder. In fact, the third pillar in the Origins Program is a concept Angel and Woolf have been considering for decades in one form or another.

"I remember thinking about the problem of looking for planets very early on, back in the 1960s," Woolf told me in his high, soft voice as we sat in the Steward Observatory conference room during my visit in January 1997. Woolf's attention drifted away from the question of planets, but it returned during the 1970s when he attended a meeting about Project Cyclops. This was a NASA proposal for a huge array of radio telescopes that would listen for signals from alien civilizations. (Cyclops was never built, but the proposal alone got many scientists excited about the search.) People wanted to know what options existed for filling in various parts of the Drake equation. Looking for planets, especially in infrared light, was one idea floating around. "I pointed out that it would be very hard to do, since planets are close to stars," Woolf recalls. "But then Ron Bracewell came up with the idea of nulling."

Ronald Bracewell is a Stanford electrical engineer, a remarkably creative scientist who has made important contributions to radio astronomy, medical imaging, and the search for extraterrestrial intelligence. What Bracewell proposed in 1975 was that you could use interferometry—at that time still mostly a theoretical technique—to make ultra-precise images of objects in space and also to eliminate objects you didn't want to look at.

Bracewell realized that interference is so sensitive a phenomenon that its effects change from one side of an image to the other. If you arrange things so that the object in the center of an image is constructively interfering with itself—a star at its very brightest, for example, as in Shao's interferometer— then objects at the edge will be interfering destructively. They'll essentially disappear.

Conversely, you can focus your interferometer so that the object in the center disappears and the things away from the center stand out. That turns out to be ideal for planet hunting. Even in infrared light, where stars are relatively dim and planets relatively bright, a star outshines its planets by a factor of ten million or more to one. Picking a planet out of such a glare would be like trying to see a candle next to a searchlight. With destructive interference, though, you can make the star disappear—you can "null" it—leaving the planets to shine all alone.

Woolf's thinking was further galvanized by a talk that Angel gave in the early eighties on the search for life. (As in all true partnerships, each of the British Arizonans has inspired the other at different points in their careers.) Woolf and Angel began talking seriously about how such a search should proceed. They concluded that for several reasons the infrared band of the spectrum was an ideal place to look: There's less "noise" from the star, and

the spectroscopic signatures of interesting molecules are most prominent in
the infrared. "All the reasons," says Woolf, "that people now accept."

The concept of nulling was so compelling to Woolf and Angel that they
eventually incorporated it into their ground-based interferometer, the LBT.
Its mirrors will be mounted at the separation that makes nulling out a star at
the center of the image easiest: fourteen meters apart, center to center, or a
bit less than six meters apart, edge to edge. The astronomers believe it's possible, though not certain, that the LBT will for the first time be able to image planets directly, as opposed to detecting them indirectly through
Doppler shifts or astrometry. It may even be able to photograph Jupiters, if
they exist, around the nearest few dozen stars. Even if it can't, the LBT will
be able to do plenty of other research that is central to the question of life on
other worlds. Like the NGST, for example, it will be able to image disks of
dust on the verge of turning into planets—though with twice the light-
collecting area, it will be more sensitive than the space-based telescope.

Nulling will also allow LBT to probe more mature solar systems to see
how dusty they are. Our own is loaded with dust, especially among the inner
planets. This dust glows brightly in the infrared part of the light spectrum,
creating what astronomers call zodiacal light, a glowing fog that might make
small, habitable planets hard to see even with a nulling interferometer. "The
question," Ed Weiler told me, "is whether our solar system is typical. Or is it
unusual in that it has a lot of dust? Or is it unusual in that it *doesn't* have a
lot? If we go out and look at a hundred stars and ninety-nine of them have
significantly more dust than we do, we can forget this whole thing." LBT—
along with the other large telescopes being built by Angel's rivals, the
NGST, and an orbiting infrared telescope scheduled for launch before the
year 2000—will help determine whether direct searches for Earth-like planets are practicable.

Zodiacal light is a major worry for the Planet Finder as well; it's nearly as
hard to see out of a fog as into one. Even in space, an infrared interferometer of the sort Bracewell had suggested would be swamped by zodiacal light.
One way to overcome the problem, Woolf told me, would be to make your
space interferometer out of truly enormous mirrors, hundreds of feet
across—though that was obviously an absurd solution.

Then Alain Léger, a French astronomer, offered what Woolf and Angel considered a brilliant suggestion. Léger noted that the zodiacal dust trails off markedly out near the orbit of Jupiter. If you sent your interferometer out there, the mirrors could be reduced to a manageable size. In the meantime, Angel worked out the optimal design for such an interferometer: a linear array consisting of two large telescopes close in and two smaller ones on the edges. In this configuration, Planet Finder would be about seventy-five meters long overall—one long structure (deployable, of course; even Angel concedes that) or perhaps four independent spacecraft flying in tight formation.

When aimed at a star, the combined light from the four telescopes will create a dark band through the center of its field of view, through which no light shows, and then alternating bands of clarity and obscuration as you look out toward the edges. If you then rotate the entire spacecraft, the zone of obscuration that goes through the center stays there—like the center of a rotating propeller—but every other part of the field of view is visible at least part of the time. Any other object that's there, emitting infrared light, will flicker on and off as the zones of clarity sweep across it. Gradually, like frames from a movie, these flickers will build into a coherent picture not only of the spacing and mass of planets in a solar system but also of their atmospheric composition.

Planet Finder could nail down the F_e element of the Drake equation by showing astronomers just how common Earth-like planets are. But Ed Weiler and Dan Goldin hope to push the equation still further. Life wreaks havoc on planetary atmospheres. Plants suck in carbon dioxide, for example, and spit out oxygen; they die and rot and release their stored carbon as methane. Both oxygen and methane are highly reactive, though; they'd quickly combine with each other unless they were continually being replenished. Planet Finder will attempt to take spectral measurements of a planet's atmosphere; if it saw lines of oxygen and methane, it would be highly likely that life is responsible.

This experiment has already been done. In 1990 the Galileo spacecraft, which is now orbiting Jupiter, had a close encounter with planet Earth. In order to get the probe all the way out to 450 miles away from the Sun, NASA sent it out to loop around Venus once and Earth twice, using the inner planets' gravity to sling Galileo toward Jupiter. On one of its Earth flybys, Galileo aimed its detectors at the planet and detected both oxygen and methane. Earth is inhabited.

At interstellar distances, unfortunately, both oxygen and methane are hard to spot; their lines are vanishingly faint. Where there's ordinary oxygen, though, there's inevitably ozone as well, a molecule made of three oxygen atoms. And ozone puts a clear, distinctive mark on a spectrum. While there is no good stand-in for methane, atmospheric chemists believe it's very unlikely that both oxygen and liquid water can exist easily on a planet at one time. So it's ozone and water, rather than oxygen and methane, that Planet Finder will be trying to detect.

While NASA isn't ready to buy Roger Angel's design for the NGST, the agency has adopted Angel and Woolf's concept for the Planet Finder. The European Space Agency has essentially done the same for its own, almost identical project, called Darwin. Neither agency even mentions the existence of a rival Planet Finder in its public relations materials because such an admission might be suicidal. The news that Darwin is being funded would probably persuade Congress to pull the plug on Planet Finder, and vice versa.

No matter which one is finally built, though, Neville Woolf and Roger Angel's mark will be on it—and on the NGST and on the ground-based interferometry and on adaptive optics and on the construction of giant telescope mirrors as well. Mayor, Queloz, Marcy, and Butler will always be recognized as the ones who started to chip away at the Drake equation. But in the end it may be Angel and Woolf who crack it wide open.

Chapter 10

WHERE DO YOU LOOK?

Given the fact that its surface temperature is hot enough to melt lead, 51 Pegasi B is not the first place you'd think of to look for alien life. That's not to say categorically that life can't exist there. It only means there's no chance of finding life as we know it.

The phrase "as we know it" is so routinely tacked on to discussions of extraterrestrial life that it seems almost like a throwaway line. It represents, however, a fundamental prejudice that Drake-equation scientists freely acknowledge. Only the foolish would insist that life as we know it is the only sort of life that could possibly exist, considering that the universe contains at least fifty billion galaxies, each containing a hundred billion stars. If science fiction authors and *Star Trek* writers can envision life as we don't know it—disembodied energy patterns, crystals that think, intelligent clouds of gas—then surely the universe is equally creative.

If that's true, then even the first element in the Drake equation—the frequency of Sun-like stars—may be too conservative. Why limit ourselves to Sun-like stars? Who's to say that radiation-eating philosophers don't live on Alex Wolszczan's pulsar planets, for example? Life as we don't know it could be a possibility around stars like the Sun as well. Perhaps monsters made of granite, with molten lead for blood, are living the good life on Peg 51 B. We can't prove they aren't.

Nevertheless, bioastronomy (or astrobiology or exobiology or Drake-equation science) is difficult enough without taking on such questions. Life in the universe might come in all sorts of forms. We only know for sure about the existence of one. It requires a certain type of chemistry; it lives within a certain range of temperatures; it alters its own environment in certain characteristic ways. Finding this sort of life at a distance of tens of light-years won't be easy. The Planet Finder is ingenious but also terrifically ambitious. Sending four satellites out to Jupiter and flying them in formation with a precision of millionths of an inch will not exactly be a piece of cake.

At least, though, life as we know it is based on known chemistry, lives under known conditions, and leaves a predictable mark on its environment. Life as we don't know it is based on unknown chemistry, lives under unknown conditions, and leaves an unknown mark on the world around it. How can you search for something like that? Until someone has a better idea, scientists are stuck with the merely overwhelming job of looking for reasonably familiar sorts of organisms. The term F_l in the Drake equation, the fraction of planets that gives rise to life, should properly be F_{lawki}, the fraction that gave rise to life as we know it.

The first important characteristic of life as we know it is that it's built mostly from compounds that contain carbon atoms. Look in a biochemistry or organic chemistry textbook. The letter C, for carbon, is all over the place—in amino acids, in proteins, in sugars, in enzymes, in DNA.

This is no accident. Living organisms couldn't extract energy from their environments, reproduce, or pass on biological information to their offspring (thus permitting them to evolve) without a huge number of complex, specialized molecules to help them do it. Because it's such a versatile, friendly atom, carbon is ideally suited to form such molecules. It combines easily with atoms of other elements, including oxygen, hydrogen, and nitrogen, to form an almost endless variety of elaborate chemical compounds. Silicon is the only other element that comes close to carbon in its ability to mate with other atoms, but silicon-based compounds are generally more unstable than their carbon equivalents. Beyond that, carbon is the fourth most abundant element in the universe, ten times as plentiful as silicon. There's plenty of it to work with. Nature couldn't possibly have made a more sensible choice.

Another requirement for life as we know it is liquid water. Without some sort of fluid medium to float around in, all these complex carbon compounds would have a hard time getting together. If they're going to exchange energy, to pass on biochemical information, to organize themselves into even more sophisticated and complicated structures—from simple molecules to complicated self-replicating molecules to cells, tissues, organs— they need to run into each other frequently. Water is ideal: It's a powerful solvent and, like carbon, is abundant all over the universe.

Much of the universe's water isn't in the right form, though. Out in deep space, water is frozen solid. In close to stars, it can exist only as vapor. Water takes useful, liquid form when its temperature is between 0 degrees and 100 degrees Centigrade, 32 degrees and 212 degrees Fahrenheit (at normal atmospheric pressure, that is; if the air pressure is high enough, water can stay liquid up to a few hundred degrees). The region where water is wet is known as the habitable zone. It has to be close, but not too close, to a star. If you see a planet through the Planet Finder, it should presumably be in the habitable zone if you expect to detect evidence of life.

Stars come in all sorts of sizes and shapes—big, bright, and hot; small, dim, and cool; or pretty much anything in between. That means habitable zones come in different sizes as well. The habitable zone for a hot star is obviously farther out than it is for a cool star. For all of the planets discovered so far, though (except Alex Wolszczan's pulsar planets), that's not an issue. Using the life-as-we-know-it argument, astronomers have concentrated their limited telescope time only on Sun-like stars.

But even for this single class of stars, judging where the habitable zone lies is a tricky process because a given star's habitable zone has to do only partly with the star itself. It has also to do with the planet. A world's surface temperature depends not just on how much heat beams in from outside but on how the planet handles that heat.

The Moon, for example, lies smack in the middle of the Sun's habitable zone, yet isn't even remotely habitable. It's small and doesn't have enough gravity to hold on to an atmosphere. Without an atmosphere, incoming solar heat can't spread evenly around its surface, so the daylight side is hot and the night side bitterly cold. Any water on the night side, or even in a shadow on the day side, will be frozen. (In fact, the Department of Defense's lunar-orbiting Clementine space probe detected what scientists think is ice sitting in permanent shadow at the bottom of a crater near the Moon's south pole. It's probably the remains of a comet that crashed ages ago.) Any ice illuminated by the Sun will melt—but then, because the Moon has no atmospheric pressure, the water will evaporate almost immediately and, like any other gas, dissipate into space.

If a planet is big enough to hold on to its atmosphere, on the other hand, the blanket of air not only spreads heat around evenly but also traps it. Incoming sunlight heats the planet; then, when the planet tries to radiate that heat back out into space, it can't. The atmosphere acts somewhat like the glass walls of a greenhouse. Thanks to years of harping by environmentalists, this greenhouse effect is widely misperceived to be an inevitably bad thing. The truth is that without the greenhouse effect, due mostly to the heat-trapping properties of carbon dioxide, the Earth would be about 11 degrees Fahrenheit colder than it is now, on average. It's only the *extra* warming from human-generated greenhouse gases that causes a problem.

If you moved the Earth closer to the Sun, the temperature would obviously rise because the Sun would be that much brighter. The rise wouldn't be linear, though. The temperature wouldn't simply increase in proportion to the extra heat. As the air grew hotter, water from streams and oceans would evaporate. The water vapor would itself act as a greenhouse gas, accelerating the warming, which would force even more water to evaporate, which in turn would boost greenhouse warming even more. Temperatures would go up like a rocket, slowly at first and then faster and faster—global warming on a scale more catastrophic than Greenpeace's worst nightmare.

Eventually, all the water would be vaporized and its molecules ripped

apart by solar ultraviolet light. Some of the hydrogen and oxygen would recombine with sulfur to produce sulfuric acid. The rest of the hydrogen
would boil off into space, leaving oxygen behind to combine with surface
rocks. If the Earth were brought in to orbit at .73 astronomical units, or
about seven-tenths of its ninety-three-million-mile current distance from the
Sun—the place where Venus currently lives—it would become a twin of
Venus: bone-dry, with a permanent stratospheric haze of sulfuric acid floating in a carbon dioxide atmosphere and an average surface temperature of
900 degrees Fahrenheit. Given that its atmosphere will act as a heat trap, an
Earth-size planet orbiting a star like the Sun has to be farther out than Venus
in order to be habitable.

That being the case, Geoff Marcy's dramatic introduction of 70 Virginis
B at the American Astronomical Society was misleading. He had said that 70
Virginis B might have liquid water on it. 51 Peg B would be too hot; 47
UMa B, as Marcy pointed out, was too cold. 70 Vir, whose cloud tops should
be hovering at about 185 degrees Fahrenheit, was "just right," though admittedly still on the tepid side. (This point was inevitably followed by a suggestion that the new planet be named "Goldilocks." Marcy wasn't prepared to
go to the International Astronomical Union, which rules on names for newly
discovered objects, with a proposal quite as silly as this. He did have a serious
name to suggest for 51 Peg B: Bellerophon, who rode the winged horse Pegasus in Greek myth. The IAU is considering it.)

But 70 Vir B is not just right. First of all, as Marcy acknowledged, it is almost certainly a gas giant like Jupiter, with no solid surface. Unless an entire
fauna were somehow able to evolve without the benefit of land or sea, the
only way it could harbor life as we know it would be indirectly, on the surface of one of its moons, should they exist. The moons wouldn't be just
right, either, however—not even close. 70 Vir B orbits its star at .43 astronomical units out, about half the orbital distance of Venus. If the moons
were large enough to hold an atmosphere (as Neptune's Titan is, for example), they'd be as hot as ovens and totally dry; if they were too small, they'd
be airless and dry.

When you take into account the effects of atmosphere, the inner edge of
the habitable zone, according to James Kasting at Pennsylvania State Uni-

versity who is considered the world's leading researcher on the topic, is nowhere near 40 million miles from its sun, where 70 Vir sits, nor 67 million miles out, where Venus is, but at 88 million—only 4.5 million miles shy of Earth's current position.

Now imagine the experiment in reverse: Attach a cable to Earth and haul it out farther from the Sun than it is now. The greenhouse effect, which worked against the planet closer in, now keeps it from freezing. You could drag Earth out to about 127 million miles from the Sun, Kasting figures, before it would get unlivably cold. Mars lies just beyond that limit, at about 151 million miles from the Sun. With an atmosphere only six one-thousandths as dense as Earth's, Mars's surface temperature hovers at around minus 60 degrees Centigrade, or minus 76 degrees Fahrenheit.

When the Planet Finder goes into orbit out in the neighborhood of Jupiter, it will be looking especially for planets that fall into the narrow range of orbital distances from their stars of between 88 and 127 million miles out. Actually, though, as Kasting explained to me during a visit to Penn State in November 1996, it's an even narrower range that's likely to be truly interesting. A planet has to be in its star's habitable zone for a reasonable length of time, he figures, if it's going to have life on it. 51 Peg B and any moons it might have would have passed through the habitable zone, for example, as it spiraled in toward its Sun, but only for a few thousand years. That hardly counts.

Even for more conventional planets, said Kasting, the habitable zone changes slowly over time because the Sun and stars like it get warmer as they age. The Sun was more than 30 percent cooler 4.5 billion years ago when the planets formed than it is today. Its habitable zone was closer in then and has been moving outward ever since. The best bet for finding life, he said, is in something called the *continuously* habitable zone—the region where the original habitable zone and the present one overlap and where conditions friendly to life have existed all along, giving sophisticated organisms long enough to evolve.

"If you look at standard planet formation models," Kasting said, "you would estimate that the average solar system has about a 50 percent chance of having a planet in the continuously habitable zone." Pick ten solar systems, and you should get five with planets in the right place. (Note that he was talking about standard solar systems, which means that most of those systems already discovered don't count.) Of those five, some will be smaller than Earth, perhaps so small that, like the Moon, they won't be able to hold on to an atmosphere. Even for a young Earth, though, the atmosphere had to come from somewhere. It couldn't be captured, the way Jupiter and the outer planets sucked in their huge atmospheres; the Earth didn't have powerful enough gravity. Besides, there wasn't much gas to capture; most of it had already been blown off toward the edges of the solar system.

Some gases, though, would have come from the formation process itself, part of the mix of elements in the original disk. More were locked within the rocky substance of the new planet. Oxygen and carbon, trapped in interstellar dust in the form of rocky silicates and microscopic diamonds, were being cooked in the new world's molten interior and spewed out through cracks in its newly forming crust. The young Earth was intensely volcanic, and each volcano belched out gases freed from the liquid rock to form a blanket of air—foul, unbreathable air, but air nevertheless—around the planet.

Then the comets began to fall. When the Sun blew the inner solar system clean of gases, some never made it past Jupiter or Saturn or Uranus or Neptune. Most did, though, and finally, when they were out at the frigid fringes of the solar system, beyond the Sun's influence, they condensed into ices—frozen water and carbon dioxide, mostly; ordinary ice and dry ice.

In a replay of what had happened in the planet-formation process, ice grains stuck together to form pebbles, which formed rocks, which formed boulders of ice, and then on up to mountains and, finally, small icy planets—including, probably, Pluto. Collisions and close encounters between the forming bodies in this mirror solar system, along with gravitational poaching by Neptune and Uranus, sent many of the chunks veering off at odd angles, including on trajectories that took them by the millions back in toward the Sun and the inner planets.

For half a billion years or more, the comets rained on the Earth, delivering one catastrophic impact after another—each as large and destructive, on average, as the one that scientists believe wiped out the dinosaurs sixty-five million years ago—except that these fell every month or so, year after year, century after century, for something like five hundred million years.

The comets delivered destruction, but they also delivered gases to beef up the atmosphere. And they delivered water by the trillions of gallons— enough comet meltwater to form an ocean a half-mile deep covering the entire globe, according to some estimates. The bombardment is still going on, although at a less disruptive pace. Observations by an Earth-orbiting satellite in 1997 suggested that tens of thousands of house-size comets strike the Earth's upper atmosphere every day of every year. They're too small to reach the surface without blowing to pieces, but they continue to add thousands of tons of water to the planet each year. By four billion years ago, five hundred million years or so after it formed, Earth was sitting right in the center of the habitable zone, with plenty of liquid water and an atmosphere to keep it that way. If Geoff Marcy had been looking through his telescope, and discovered Earth, he would have pronounced the new planet "just right."

The one ingredient remaining to get life started on the primordial Earth was a suite of complex, carbon-based compounds. But here, too, comets could have helped. The vast interstellar clouds that give birth to stars and planets are made mostly of hydrogen, but they contain plenty of other elements, including atoms of oxygen, silicon, nitrogen, and carbon. Huddled together on minuscule grains of silicate and carbon dust, and bathed in ultraviolet light from nearby young stars, these atoms combine and recombine over millions of years into an enormous variety of complicated molecules.

Astronomers know this is true because every molecule has its own unique way of vibrating when plucked by electromagnetic radiation from nearby stars, a song it sings in the radio part of the spectrum. In thirty years of listening, radio astronomers have heard the songs of alcohol, ammonia, hydrogen cyanide, methane, formaldehyde, and more than five hundred other chemicals, many of them highly complex, many of them organic.

These organic chemicals would have come along in the general collapse of our own interstellar cloud, and we know for a fact that at least some of them survived. The comets are mostly ice, but they're contaminated with primordial dust as well. The dust blows off when they swing through the inner solar system, which they still do in a feeble echo of that early, terrible bombardment. As Halley's and Hyakutake and Hale-Bopp have breezed by in the past decade, astronomers have scanned the dust and found methane, ammonia, and, during Hale-Bopp's close approach in the spring of 1997, formic acid as well. This last is especially interesting because it's a direct precursor of amino acids—you just have to add ammonia, which is conveniently present. Evidently this has already been happening. Astronomers found less formic acid in Hale-Bopp than they expected. Some of it has gone somewhere. If it's into amino acids, as comet experts believe, their signature is too faint to detect yet. But these carbon-based chemicals that act as the building blocks of proteins have been extracted intact from meteorites that have fallen to Earth, most famously from the Murchison meteorite that fell in Australia in 1969.

This suggests a second route by which organics came flying in from space. Meteorites are simply chunks broken off asteroids, which are themselves left-over fragments from the solar system's formation. Asteroids can fall all in one piece, too. The mile-wide Meteor Crater in central Arizona was caused by an asteroid a few hundred feet across, crashing to Earth tens of thousands of years ago. (The asteroid or comet that wiped the dinosaurs from the face of the Earth sixty-five million years ago was a few miles across and gouged a crater one hundred miles wide off the Yucatan Peninsula in Mexico.)

In geologic terms, these impacts happen all the time, and they'll continue to happen. During the past several years, telescopic searches have found a handful of previously unknown small asteroids zipping by uncomfortably close to the Earth. There are undoubtedly more, and sooner or later one won't miss. Astronomers periodically lobby Congress for money to build an asteroid-spotting network so we can at least identify the dangerous ones, but without much success so far.

Back during the solar system's early days, there were probably many more of these loose asteroids flying around, left over from the just-completed assembly of the planets. So while the Earth was being bombarded with comets during its few hundred million years of life, it was simultaneously being pelted with asteroids. Both would have left terrible scars on the planet, but the craters from that double bombardment have long eroded away or been carried down into the interior of the planet by the forces of plate tectonics.

On the Moon, though, the craters remain. The dating of crater rocks brought back by the Apollo astronauts shows that the bombardment with both types of heavenly missiles continued for about seven hundred million years, until about 3.8 billion years ago.

The assault stopped only when the ammunition ran out. Whatever comets and asteroids were left had either settled into stable, nonthreatening orbits for the most part or had been flung to the outer edges of the solar system by Jupiter's dominant gravity. (The role of Jupiter in ending the bombardment was first recognized by George Wetherill in the 1970s. By clearing the inner solar system of dangerous missiles, Jupiter deserves part of the credit for permitting life to get started. Without a Jupiter in about Jupiter's position, other solar systems might not be so lucky.)

Although the rain of mountains made Earth a dangerous place while it was going on, it brought in organic chemicals that might have given life on Earth a head start. High-speed impacts—above about eight kilometers per second—would probably have fried any interesting chemicals on board the crashing objects. It's almost certain, though, that the impacts came in a wide range of velocities. Some of the objects would have been essentially catching up with Earth in its orbit and would have come at relatively low speeds, allowing them to survive chemically intact.

Organic chemicals might also have come to Earth by a less violent route as well. Even today, long after the solar system has calmed down, comets shed plenty of dust as they fly through the inner solar system. Asteroids are sturdier, but they create dust as they crack together and pulverize one another. The zodiacal dust that will force Planet Finder out to Jupiter is comet and asteroid dust. Each year tons of the stuff filters down onto Earth's surface—you can collect it from the rain gutters on your house. Four billion years ago there were more asteroids, more comets, and correspondingly more dust, with more filtering to Earth. And much of it, then as now, was presumably organic.

Or maybe, despite the ample supply of organic matter out in space, Earth made its own. In the early 1950s a young graduate student named Stanley Miller did an experiment at the University of Chicago to try to reproduce conditions on the young Earth. He filled a test tube with water to simulate the ancient ocean and added what scientists then believed was the planet's primitive atmosphere: methane, ammonia, hydrogen, and water vapor. Then Miller sealed the whole thing up and ran an electric spark through it—lightning in miniature.

After about three days, Miller found that the water in his apparatus had turned pink and murky—contaminated, it turned out, with amino acids and other organic molecules. The so-called Miller-Urey experiment (the Nobel Prize–winning physicist Harold Urey was Miller's advisor) seemed to show that the Earth itself, without outside help, could have served as its own chemical lab.

Geochemistry has since caught up with Miller. The current consensus is that Miller's atmosphere wasn't really like the primitive Earth's: too much methane, too little carbon dioxide. Miller himself isn't so sure. He is still pursuing active research as one of five senior scientists who comprise the NASA Specialized Center of Research and Training in Exobiology (NSCORT) at the University of California, San Diego (another NASA effort that will probably be absorbed under the Origins rubric).

As far as geochemists are concerned, the Miller-Urey experiment is of historical interest only. As far as Miller is concerned, they're overstating the case. "Let's state the facts," he told me as I sat in his office during a visit to the university's seaside campus in December 1996. "There's no evidence for one atmosphere on the early Earth over another. Calculations of model atmospheres are not evidence no matter how meritorious they are. I'm very adamant about that."

Miller's was the first plausible mechanism to explain how Earth might have acted as its own natural organic chemistry lab. Lately, scientists have come up with another. This new theory suggests that the chemistry of life originated not in space or from the primitive earthly atmosphere but from volcanic holes in the ocean floor. These still exist today—cracks in the bottom that belch superheated water, hydrogen sulfide, carbon monoxide, and dissolved metals. Deep-ocean vents would have been much more common on the young, hot Earth, whose crust had just started to cool. Recent experiments by a German patent attorney named Günter Wächtershäuser show that organic molecules can form under these conditions as well, using crystals of iron pyrite—fool's gold—as a sort of scaffolding around which to assemble themselves. (Wächtershäuser isn't paid for his science, so he's technically an amateur, but he does have a Ph.D. in biochemistry.)

Each of these scenarios is plausible; all of them may even be true, pieces of a multistage process whereby life began. What they mean for Drake-equation science is that organic chemistry—that is, complex, carbon-based chemistry—is something Nature indulges in every chance it gets. Organic chemistry happens in deep interstellar space. It happens, evidently, in comets. It happens when you zap a few simple chemicals with a jolt of electricity. It happens when you charge seawater with volcanic fumes. It happens, in short, under all sorts of completely different conditions.

That being the case, it's hard to imagine how organic chemicals can avoid being present on the surfaces or in the oceans of worlds in faraway solar systems. Interstellar dust clouds are the universal raw materials for planets. Comets and asteroids are almost certainly universal as well; unless theorists are kidding themselves badly, you can't form a solar system without making both.

Observations of nearby stars seem to bear this out. Several mature stars, notably Beta Pictoris, have disks of dust whirling around them (Beta Pictoris's dust disk was the first one ever photographed). These can't be pre-planetary disks because the stars are too old for that. They can't just be dust left over from the star's formation billions of years earlier, since all the dust would long since have been swept up into larger objects. They must be fresh comet and asteroid dust.

Finally, young planets in the habitable zone will inevitably have comet-delivered water on them and volcanic activity. Their atmospheres will contain some combination of hydrogen, oxygen, carbon, and nitrogen. One way or another, organic chemistry is something you probably can't escape.

The organic chemicals in interstellar clouds, asteroids, and comets, having no access to liquid water, won't organize themselves into living organisms, as far as anyone knows. (The British astronomer Fred Hoyle and the Sri Lankan astronomer Chandra Wickramasinghe disagree. They argue that infrared emissions given off by interstellar dust grains are consistent with the notion that bacteria are floating around in space. Their colleagues are extremely skeptical of this interpretation.)

On the surfaces of planets, though, perhaps life is common. With liquid water, an atmosphere, and organic chemicals, maybe it's even inevitable. And that raises the question of Mars, a question that has been dormant for decades. The first space probes that flew past the red planet in the 1960s destroyed all hope that any sort of complex life form existed there. It was evidently a dry, inhospitable place—a desert planet.

The Viking orbiters that photographed the planet in detail a decade later, however, reignited that hope, not so much for the present but for the past. Mars was indeed dry and pockmarked with craters, but ancient river valleys meandered unmistakably across the red dusty landscape, branching and twisting just as they do on Earth. The surface bore deep channels scoured by catastrophic outflows of water, perhaps freed when huge dams of ice gave way. Telltale evidence of long-dry lake beds marked the land. Water had flowed across Mars at least 3.5 billion years ago, which means the planet must have had a substantial atmosphere as well.

According to the theory of habitable zones, though, that doesn't make sense. Mars is too far from the Sun. "It's a real puzzle," James Kasting admitted to me, "especially considering that the Sun was significantly cooler then. Something was keeping Mars warm." It may have been, he speculates, some additional heat-trapping gas besides carbon dioxide—methane, for example—or it may have been that carbon dioxide ice clouds amplified the effect of gaseous CO_2 and trapped even more heat than you'd assume.

Theory will presumably catch up to fact at some point. The fact is that Mars was warm and wet. It was bombarded with organics in comets and asteroids, dusted with organics, and perhaps created its own organics. If life is inevitable or even likely, then there is every chance that life began on Mars as well.

It didn't have all that much time to develop, however. The same comets and asteroids that delivered water and organics to Mars, as they had to Earth, and that blasted the craters into its rusty surface, sent great gobs of its atmosphere flying off into space as well. Only half as massive as Earth and with half its surface gravity, Mars couldn't hold on to its air; the atmosphere was gradually stripped away in a process known as "impact erosion."

Mars's smaller size meant its atmosphere was doomed for another reason.

On Earth the thick layer of molten subsurface rock known as the mantle continues to churn slowly, like a thick oatmeal on a stove. The crust on top flows with it—cooling from molten magma to form broad crustal plates where the mantle bubbles upward, flowing across the surface like a huge conveyor belt and then disappearing to melt again millions of years later. Among other things, this process recycles the atmosphere continuously: Carbon dioxide is washed out of the air by rain, and reacts with minerals to form carbonate rock. The rock eventually slides down into the mantle where it melts, releasing the carbon for volcanoes to deliver into the air again—essentially the process that created the atmosphere in the first place, recapitulated.

On small, fast-cooling Mars, though, plate tectonics stopped long ago. The planet has the largest, tallest volcanoes in the solar system. The biggest, Olympus Mons, is 15 miles high and 450 miles across. But those volcanoes are eons dead, with no molten rock deep underground to well up and replace them. What was left of Mars's carbon dioxide after the comets and asteroids ripped much of the atmosphere away bound itself into rock long ago, never to escape. Without an atmosphere, the water disappeared from the surface, and so, presumably, did whatever life that had arisen.

From the point of view of the Drake equation, though, evidence that life arose on Mars—a second genesis within a single solar system—would be extraordinary. It would suggest that F_l may not be one, the optimistic number Drake and the others assigned it at the original 1961 conference on alien life, but two or more.

Life on Mars would have had to overcome one major obstacle, however: It had very little time to get started. The early bombardment of asteroids didn't finally let up until 3.8 billion years ago. The water disappeared 3.5 billion years ago. That doesn't sound like much time for so profound a transformation, for the leap from non-life to life. If an additional requirement for life—after a planet to live on, water, and carbon compounds—is an uninterrupted billion or more years of time, then the chance of finding life elsewhere may be small after all.

Chapter 11

LIFE IN THE SOLAR SYSTEM

The lump of rock sitting in my palm didn't look like anything special. It was a little heavier than I would have expected, given its size, but a reddish overtone to its otherwise gray-black color explained that easily: The rock contained a lot of iron ore. That was all that was apparent to my uneducated eye.

The young man who had just handed it to me knew better. Steven Mojzsis—tall, almost always serious-looking, but with a deadpan sense of humor that could strike without warning—is a postdoctoral student in the NSCORT exobiology program at the seaside Scripps Institution of Oceanography at the University of California, San Diego. As Drake-equation scientists have realized since the beginning, to estimate the general probability of life arising elsewhere we must understand how it happened here on Earth—whether this planet beat the odds or surrendered to them.

To that end, Mojzsis had been studying this piece of stone ever since he'd chipped it out of a hillside in Greenland months earlier—not only with his own eye but also with plenty of sophisticated laboratory equipment, including an electron microscope and an ion microprobe. With the help of two other colleagues and his advisor, Mojzsis had gradually unraveled the rock's rather remarkable history. (The advisor, Gustav Arrhenius, is one of NSCORT's senior scientists, along with Stanley Miller. He is also the grandson of the Swedish chemist Svante Arrhenius, who was one of the first scientists to describe the greenhouse effect and who also proposed at the turn of the last century that life could have arrived on Earth fully developed rather than starting out here. This was the original panspermia theory, which in modernized form Fred Hoyle, Chandra Wickramasinghe, and a very few others still support.)

Thousands of millions of years ago, went Mojzsis's story, dust and debris settled slowly down to the bottom of a primordial sea. The detritus collected there in layers, fractions of a millimeter at a time pressing down on the sediments that preceded them, and pressed down in turn by what came after. As they were buried deeper and deeper, the bottom layers of sediment were finally squeezed into rock—sedimentary rock, one of the three types that elementary school kids learn about in their science classes.

Then, agonizingly slowly, the rock was sucked deep underground. The huge sections of Earth's crust known as tectonic plates were then, as they are today, in constant motion, emerging from semi-molten rock as it oozed to the surface from many tens of miles down, sliding across the surface like thousand-mile-wide conveyor belts and disappearing down again into enormous cracks called subduction zones. The entire Japanese archipelago currently sits at one edge of such a subduction zone. The seafloor of the western Pacific is sliding slowly down into the Earth off the coast, triggering frequent earthquakes and fueling active volcanoes as it reluctantly goes.

The same thing happened to that ancient sedimentary rock. It slid across the surface of the Earth and plunged down, breaking up and partially melting at subterranean temperatures of 500 degrees Centigrade. Its fragments were mixed into the magma below as raisins are folded into cake batter. And then, ages later, it oozed out of the Earth again, incorporated into new crustal material. This time, though, the crust got stalled. Modern Greenland sits at the center of a crustal plate, and it has never been subducted since that last emergence 3.86 billion years ago.

What Mojzsis and Arrhenius found trapped in the heavy lump of reddish black iron-rich stone in my hand was evidence suggesting that life of some sort had flourished in that long-vanished sea. Some of the particles that floated down through its waters contained carbon—carbon bearing the tell-tale chemical signature of having once been part of a living organism. "It's circumstantial evidence," Mojzsis said. "We weren't around, and maybe this form of carbon came about through a process nobody's ever thought of." The more likely explanation, in his and Arrhenius's opinion, is that life existed on Earth 3.86 billion years ago.

Until Mojzsis chipped out his rock, the most ancient life-forms known to have existed on Earth were pillowy blobs of stone called stromatolites. These are the petrified remains of dense colonies of aquatic bacteria, and they date back to 3.5 billion years at the earliest. This is already an impressive statistic. Bacteria are single-celled organisms, but cells are already pretty sophisticated things, with a complex architecture that includes cell walls, nuclei with DNA locked inside, and energy-processing structures. These all had to evolve, which presumably takes time. A 3.5-billion-year-old bacterium implies that life had already been around for a while.

In the building next to where Mojzsis had handed me his precious sample on a warm afternoon in December 1997, Jeffrey Bada—NSCORT senior scientist number three—would, an hour or so later, explain to me how important the original discovery of stromatolites had been. Bada is a compact, trim, gray-bearded man with a relaxed, almost folksy way of speaking. (For some reason I couldn't stop visualizing Bada, dressed in the uniform of a National Park Ranger, explaining in a friendly but authoritative way how to avoid being eaten by a grizzly bear.)

Bada was a graduate student of Stanley Miller's during the 1960s, he told me, and Harold Urey was still around back then. "I was weaned on all of this stuff," he said. At the time, explained Bada, everyone knew that the Earth was 4.5 billion years old. The earliest evidence of life was microscopic fossils found in 3-billion-year-old rocks. That meant life had had 1.5 billion years to get started.

"Okay," he continued, "so how has that changed? Well, we now know that first the several hundred million years of the Earth's history was an extremely violent time. Asteroids and comets were just plastering the place." The window of opportunity in which life had to arise isn't 1.5 billion years anymore but hundreds of millions of years. "And now," he continued, "Gustav and Steve are claiming they have evidence for life at 3.86 billion, which would narrow it down almost to nothing." If Mojzsis and Arrhenius are correct in their interpretation of what's in the heavy Greenland rock, life arose at the very first moment—the very instant, geologically speaking—when it could possibly do so.

If life were some sort of highly improbable event, like winning the lottery, then the average planet should have to wait a long time before it happened. Maybe Earth just happened to be that one-in-a-billion planet that bought the winning ticket the first time out. It's much more likely, though, according to the laws of probability, that there are winning tickets in circulation all over the place, that life is highly likely to happen, given the right conditions, and to happen rapidly. If so, then the idea that life arose on Mars is suddenly not so crazy. The planet's window of opportunity may have been brief—only a few scores of millions of years, perhaps, between the end of the bombardment and the loss of atmosphere. But it may have been sufficient.

The evidently quick appearance of life on Earth also suggests that life might have happened more than once here. Although the rain of comets and asteroids was probably more or less constant, the "less" might have included lulls lasting tens of millions of years. It's not at all implausible to imagine that life began, then got wiped out, then began again, then got blasted into oblivion—a half-dozen times, maybe, or more—before the bombardment ended.

Conversely, it's also possible that life could have lasted through a few gigatons' worth of comet blasts. This presupposes that the first living cells did not arise, as Charles Darwin thought they did, in "a warm little pond" or, indeed, anywhere on or near the surface of the planet. The energy from constant bombardment would have remelted much of the land surface and boiled off the first few hundred feet of ocean.

Perhaps, though, life came only late to the surface of the Earth. As the work of Günter Wächtershäuser proves, the organic molecules that evolved into life may first have swirled into existence around hot volcanic vents that cracked the floors of the young Earth's oceans. Not only that but Wächters-häuser's most recent experiments, reported in 1996, showed that the hot, wet, sulfurous conditions surrounding the vents could also give rise to mole-cules called thio esters, which are structurally related to the energy-processing molecules in living cells. Wächtershäuser's work suggests that metabolism may have evolved first and that cells came along later to take ad-vantage of this ready-made mechanism.

More broadly, his experiments suggest that life may have originated in a place pretty well insulated from much of the devastation going on above sea level. The first life-forms on Earth may not have lived in tidal pools or estu-aries but in natural bomb shelters. Hidden under an insulating blanket of water, these most ancient of cells would have been indifferent to the de-struction going on above. And unlike many modern deep-sea-dwelling crea-tures, they wouldn't necessarily have depended on particles of food drifting down from shallower depths to keep them alive. Their survival might have depended on complete independence from the surface. After all, the comet or asteroid that presumably killed the dinosaurs didn't melt them or inciner-ate them; it wiped out their food supply. They died relatively slowly as a pall of dust thrown up into the stratosphere by the impact hid the Sun, plunged the planet into a temporary freeze, and killed off many of the plants at the bottom of the dinosaurs' food chain.

Oceanographers learned twenty years ago, however, that some life-forms don't depend on the Sun even indirectly for their food. Diving down to the deep ocean bottom in the submersible craft *Alvin* in the late 1970s, they dis-covered what they called "black smokers"—cracks in the seafloor from which hot water dark with volcanic gases was spewing. Squeezed by the pres-sure of two vertical miles of ocean, the hot water couldn't boil, although it was coming out at well over 212 degrees Fahrenheit. These were, as far as anyone could tell, precisely the same sorts of vents that must have covered the seafloor billions of years ago.

To the scientists' astonishment, the black smokers were surrounded by liv-ing creatures—long, tube-shaped worms, sightless fish, blind crabs. The ani-mals survive, ultimately, on a rich soup of bacteria, and the bacteria, in turn, live happily on the super-heated, noxious mess spewing out of the vents.

They're chemosynthetic, not photosynthetic, converting chemicals into useful energy. They have no use for sunlight. If the Sun went out tomorrow, they would neither know nor care for decades. Since the discovery of the original black smokers, scientists have found similar communities around deep vents all over the world.

Biologists have also found "extremophiles"—bacteria that thrive on extremely hot, high-pressure conditions—in other places as well. They've been hauled up from their homes in solid rock as much as two miles underground, where temperatures reach nearly 170 degrees Fahrenheit, and from hundreds of feet below the seafloor as well. These heat-loving, chemical-sucking bacteria are extremely primitive. Analysis of their DNA proves that they belong to the group Archaea, a form of unicellular life that branched off eons ago from the line that includes modern bacteria. Or, rather, it proves that the two split; it's equally plausible that things happened the other way. It may be that modern organisms branched off from the Archaea, that the Archaea thriving above the boiling point on noxious chemicals near ocean vents and in solid rock and in the hot springs of Yellowstone National Park are the true originals. The most current findings of evolutionary biology, biochemistry, and planetary science are all consistent with the proposition that life began at the mouths of undersea volcanoes, under the harshest of conditions, and only later developed a taste for air and sunlight and mild temperatures.

If that's true, life might not have arisen in a rush after all. It might have established itself leisurely, blissfully unconcerned with the heavy bombardment at the surface. In one sense, this theory undercuts the proposition that life evolves so quickly that it's going to happen everywhere. At the same time, though, it points to a conclusion that's equally life-affirming: The habitable zone, even one broadened to include a place like young Mars, may need to be broadened still further. A planet in the conventionally defined habitable zone may be appropriate if you're looking for evidence of surface-dwelling, photosynthetic organisms, which is all the Planet Finder will be capable of doing. But even if you restrict the search to life as we know it—carbon-based, water-loving—there are suddenly a lot more places where it might be happening.

That's true even in our own solar system. Take Mars. The Viking landers that set down on the planet in 1976 confirmed what the Mariner photographs of a few years earlier had seemed to indicate: Life doesn't exist on the dry, nearly airless surface of this barren planet. Although scientists prayed it would turn out otherwise, three independent experiments revealed plenty of chemistry in the Martian soil but no biology.

But if subsea and subterranean bacteria can thrive on Earth deep below the surface, why couldn't they do the same on Mars? If life really did arise there 3.5 billion years ago and it arose, as on Earth, in a hot, Sun-less environment at the bottom of an ocean, then it needn't have died out when the atmosphere vanished and the rivers and oceans evaporated.

Planetary scientists believe that while water vanished from the surface, plenty still exists on Mars, trapped deep underground where it seeped ages ago—in aquifers, for example, like the underground reservoirs that exist all over the Earth, or simply soaked deep into the Martian soil, like dishwater into a sponge. A mile or two underground, insulated by the thick blanket of rock above, the water is kept liquid by lingering internal heat and crushing pressure, a nurturing environment where Martian extremophiles would be quite happy.

They might under other circumstances still be thriving on Venus as well; before its greenhouse effect kicked in, Venus, too, would have had plenty of liquid water, organic matter, and volcanic vents to get life started, and rocks to retreat into when the oceans boiled off. But plate tectonics probably went on much longer there than it did on Mars. Any organism that retreated to the rock would have had to re-retreat to the oceans or somewhere else when the rock slid into the planet's molten mantle. On Venus there would have been no place to hide.

Other than the Earth and Mars, then, no place in the solar system has or had liquid water in any quantity—and Mars is so cold and airless that water, if it exists at all, could survive only deep underground. The next planet out is Jupiter, five astronomical units, or 450 million miles, away. No chance of water out there, surely.

That's what planetary scientists believed before 1979, anyway. In that year the outer-planet probes Voyager 1 and 2 flew through the Jovian system.

David Black was head of the Theoretical Studies Division at Ames Research Center back then, where he had spent the 1970s sponsoring a number of summer workshops on Drake-equation science. At the Lunar and Planetary Institute in Houston in November 1996, he recalled an incident that happened just before the flyby.

"Three of our scientists—Pat Cassen, Stan Peale, and Ray Reynolds—came into my office with a dilemma," he said. They'd been doing some calculations on how Jupiter's four biggest moons—known as the Galilean moons because they'd first been seen by Galileo in 1610—would be affected by gravity. The moons were so close to Jupiter and to each other, the trio figured, that they'd experience powerful tidal forces. They'd be alternately squeezed and stretched by the complex and always changing gravitational fields they lived under. The same thing happens on Earth, and not just in the oceans. Lunar gravity causes tiny but measurable tides in the land as well, making it rise and fall by fractions of an inch.

On Jupiter's moons, under the influence of the biggest planet in the solar system, the flexing would be much greater. It would create enough internal friction to keep the moons' interiors hot. "The three came into my office," recalls Black, "and said, 'We have this idea that maybe these moons could be volcanic. But we're not sure if we should say anything in this paper we're writing.' I told them, 'You guys are nuts. Definitely say it. If you're right, you're predicting it just before the spacecraft gets there. If you're wrong, no one's going to remember a year from now.' " So they went ahead and did it.

A month later, Voyager flew by, snapping pictures furiously as it went. And there, peeking out over the edge of the moon Io, was the plume of a volcano, caught right in the act of erupting. "They won the Pierce Prize from the American Astronomical Society for that paper. Never shared the money with me, come to think of it," said Black.

If Io was hot inside, so were Ganymede and Callisto, and so, too, was Europa. That seemed to explain why Europa looked different from just about every other moon anyone had ever seen. Astronomers already knew, from looking through Earth-based telescopes and seeing in detail how the Sun's light reflected from it, that the moon's whitish surface was made mostly of ice. Now they could see that it was almost perfectly smooth as well. No craters to speak of. Instead, Europa was crisscrossed with a series of dark lines as though the surface was riddled with cracks.

Often enough in science, you get a series of observations that don't really add up. A couple of years ago, for example, the Hubble Space Telescope was able for the first time to distinguish an individual Cepheid variable star in a constellation tens of millions of light-years from Earth. The need to make precisely such an observation has been long considered one of the most important goals of observational astronomy; in fact, it was offered as one of the most important reasons to build the Hubble. Cepheid variables are like cosmic signposts that allow astronomers to measure enormous distances with high precision. Spotting a Cepheid in the galaxy M100 allowed astronomers to gauge the distance to the Virgo cluster, M100's cosmic home, with unprecedented accuracy. That led them in turn to a newly precise measurement of the size of the universe and (in a process identical to the one that high school students use to solve algebra word problems) to estimate the age of the universe as well.

The age they came up with was between eight and twelve billion years. Unfortunately, they already knew that some stars are fourteen billion years old. Each observation by itself seemed valid, but taken together they suggested the crazy conclusion that stars existed before the universe did. Something was wrong. Either the stars were younger than people thought *or* there was something wrong with astronomers' reasoning about the age of the universe *or* the big bang theory was incorrect *or* there was some sort of hitherto undetected force that was somehow distorting the measurements. Whatever the explanation, the observations had confused rather than settled matters. (It later turned out that some assumptions about measuring distances with Cepheids were probably wrong and that the universe was a shade older; it also turned out that the ages of stars had been overestimated.)

With Europa, in contrast, all the observations—white surface, dark lines, no craters, a high probability of central heating—seemed to point to a single consistent conclusion. Everything could be neatly explained if the moon's tidally induced heat was seeping out from the interior, warming the icy surface from below and melting its underside. Europa's gleaming surface was pretty clearly a moon-girdling blanket of ice, floating on water just as Earth's permanently frozen Arctic ice cap floats on the surface of the Arctic Ocean. If this explanation was correct, then the lines were cracks—places where ocean tides had ruptured the ice, letting water, or perhaps something more like slush, squeeze out and resurface on the ice, like a Zamboni machine smoothing a hockey rink. Any impact crater—for Europa is hardly immune to the asteroids and comets that periodically fall on every major body in the solar system—will sooner or later be filled in and refrozen. (Two of the other Galilean moons, Callisto and Ganymede, also have surfaces made mostly of ice. They're covered with craters, though, suggesting that they don't have much if any liquid water available for resurfacing.)

Voyager never got much closer to Europa than one hundred thousand miles away—about half the distance from the Earth to the Moon—and therefore couldn't capture details any smaller than a mile or so across. So while the theory that water lay under the ice seemed plausible, it could still be wrong. For example, scientists saw some dark patches that they couldn't really make out. Maybe they were old craters, which might mean that the surface wasn't nearly as crater-free as the observers thought. To get stronger evidence they'd have to get a much closer look.

It would take them about eighteen years to get it. In the fall of 1996, after four years of interplanetary travel, the Galileo space probe arrived in the Jovian system, the first spacecraft to return since Voyager. Among other achievements, Galileo swept at least twice to within less than four hundred miles of Europa's surface, close enough to pick out features as small as a few tens of feet.

The first pass had been scheduled for December 1996. It might seem odd, therefore, that planetary scientists chose to hold a conference to discuss the prospect of oceans on Europa in November, a month before any closeup pictures were beamed back. Actually, it made sense. The idea was to get a lot of people together from different fields—oceanographers, experts on ice, marine biologists, planetary scientists—and talk about what kind of information they were hoping to get. It was a way to prepare for the encounter.

If the Galileo encounter had happened a few years earlier, all of the scientists involved in the project would have been physically present in the Jet Propulsion Laboratory control room, waiting anxiously for the satellite to report in. You've seen it on TV a dozen times—the room full of computer monitors, the anxious-looking technicians with shirt pockets crammed full of pens, the cheers that inevitably erupt when the long-awaited images finally appear. Anyone not physically present would have to wait to examine the pictures.

Thanks to high-speed Internet access, it doesn't happen that way anymore. Almost as soon as the photos from Europa came up on JPL monitors, they came up on computers at universities around the world as well. Not only did scientists avoid many thousands of miles of travel, but they could also bring their students into these virtual control rooms to experience cutting-edge science directly.

During the pre-encounter conference, scientists pretty much agreed that the water-on-ice model for Europa's internal structure was still most probably the right one. The closeups confirmed it. But seeing what they'd roughly expected didn't make the photographs anticlimactic. "We were simply dumbfounded," says James Head, a planetary scientist at Brown University. "The images were so incredibly crystal clear, we couldn't have imagined what it would actually look like. It was as though you'd been studying leaves for all your life using a magnifying glass, and suddenly someone hands you a microscope. It was like, 'Whoa!'"

As their initial astonishment faded, the questions started to kick in. How many craters can you see? Are those things really icebergs, or could they be something else? Can we think of anything else that might create a landscape like this other than water under the ice? Head and the others began madly trying to figure it all out.

They decided there had to be water under the ice. From close up, many of the cracks looked almost precisely like those found in the Arctic ice pack on Earth, with long, parallel ridges running down their centers where the ice had been thrust upward in collisions between two gigantic thin plates.

And, yes, those really were icebergs—sharp-edged slabs of ice several miles across. They had clearly broken from the edges of larger plates, floated into new positions on a temporary sea of water or slush, and then been cemented in place by the refrozen sea. You could see on some of the bergs the marks of older cracks—ancient scars that matched perfectly the scars on the main ice sheet but canted at acute angles thanks to the bergs' motion. From the lengths of shadows they cast on the rock-hard sea, planetary scientists estimated that the icebergs towered perhaps one hundred meters above the surface. Given that icebergs show only 10 percent of their bulk above the water line, that suggests the bergs, and therefore the ice sheets they once belonged to, may be only a half-mile thick.

Their sharpened vision also gave scientists a much better idea of the number of meteor craters on Europa. They realized that the dark areas probably were not old craters after all. The scientists also realized that some of the craters they could see were secondary; that is, they weren't caused by incoming asteroids but by chunks of Europa itself falling back as shrapnel from a primary asteroid impact. It was clear that they'd been overestimating the amount of cratering on the moon and that the last resurfacing of Europa had happened even more recently than anybody had realized.

How recently? Based on the number of craters they counted, some scientists calculated an estimate of a million years. But that guess assumes the rate of cratering on Europa is similar to that on the Moon, whose history we understand pretty well. Several hundred million miles farther out and sitting in Jupiter's shadow, Europa's impact history may be very different. No one doubts, though, that the surface is, in geologic terms, extraordinarily young.

No one doubts, either, that Europa's icy surface conceals water somewhere below. It is certain as well that the same comets and asteroids that peppered Mars and the Earth and Venus with organic chemicals struck here, too. The dark patches visible in the new images may be organic comet dust left from a relatively recent low-speed impact. Over the life of the solar system, some larger objects must have penetrated the thin ice sheet to seed the ocean below with organics. Smaller comets and asteroids would have left organics on the surface—dark patches like the ones Galileo saw—and these could have percolated down to the ocean more slowly, during brief periods when new cracks exposed the subsurface waters. The comet dust sitting on Europa's surface today could be waiting for the next such opening.

Finally, it's reasonable to suspect that the moon's hot interior is forcing hot mineral-rich gases out through volcanic vents at the bottom of the Europan ocean. "What we're seeing on Europa," said Jeff Bada during my visit, which took place just after the encounter, "is basically sending chills up my spine. We have what could be a prebiotic soup going in the solar system — not four billion years ago but right now."

If the soup has been cooking for any length of time, then it may have passed the prebiotic stage. "We are *not* saying we've found life on Europa," insists Richard Terrile, a planetary scientist at the Jet Propulsion Lab who worked on Galileo, "despite some media reports that suggest the contrary." Europa is, however, at least superficially similar to what the young Earth must have been like. "It's a place that's just screaming for us to come explore," says Terrile.

The first step toward that exploration was taken soon after Galileo's December flyby. Originally, the spacecraft's mission had been scheduled to end in December 1997. Now, thanks to the startling pictures from Europa, it had been extended for two years so that the probe could return for as many as eight additional passes at the icy moon. That would give Galileo an opportunity to photograph about 5 percent of Europa's surface at high resolution — too little to let planetary scientists make broad statements about the moon as a whole but much better than the 1 percent originally scheduled.

Still, NASA will have to return if it hopes to answer a number of important questions: Are there icebergs all over the surface? Is the terrain that has already been photographed younger than average, or older? Is the ice thinner in some places than others? And is the ocean truly global?

Terrile, Head, and the others are already plotting to return to Europa in a few years to start answering these questions. Hewing to NASA's new policy of launching quick, economical missions, the Europa campaign won't involve big, expensive probes (like Galileo) that try to answer a whole series of questions all at once. It will happen in stages. The first probe will have the task of mapping the entire surface thoroughly with high-resolution cameras to get a real understanding of the geology, the forces that are shaping the surface. Mapping will also help scientists identify potential landing sites for future missions.

A radar transmitter will probe the ice from orbit to measure its thickness. If the ocean is ever going to be penetrated deeply — the obvious ultimate goal — the place to start is where the ice is as thin as possible. Radar will also be able to gauge how much the surface rises and falls under the tidal and gravitational forces of Jupiter and the other Galilean moons. The answer will help planetary scientists understand whether there's slush or water under the ice, whether there's an ocean underlying the entire surface or just pockets of liquid, and how much energy is being pumped into Europa from outside. The first probe might also try to map the surface thermally, to look for warmer spots that indicate thin ice.

This first mission, JPL scientists hope, can be launched by 2002. The second, a couple of years later, will include a probe that will land on the surface. "We think there are organics on the surface," said Terrile, "but you have to remember that this is a high-radiation environment." Jupiter's magnetic field traps charged particles from the Sun in huge radiation belts. The result is a bath of ionizing radiation at Europa's surface that's about four thousand times greater than what it would take to kill a human. Interesting organic molecules don't stay interesting very long under such conditions. The lander will therefore carry a drill, letting it bore down a meter or so to sample material that has been shielded by ice and thus undisturbed. Terrile wants to extract a sample of ice, melt it, and characterize whatever impurities lie inside.

Ultimately, if the ocean is really there—the evidence for it, while strong, is still indirect—then the next step would be to drive a probe deep into the crust. With any luck it can go deep enough to launch a tiny submarine beneath the surface. The prospect of a robotic sub cruising beneath the Europan ice cap, looking for signs of life, sounds like science fiction, but Terrile insists that it poses no obvious technological obstacles.

Naturally, Terrile and the other JPL investigators would like to get some practice at this sort of thing before they try it by remote control from 350 million miles away. As it happens, they have what looks like a perfect opportunity. Recently, Russian geologists working in Antarctica discovered that underneath the continent's thick ice cap, about four kilometers straight down, a layer of water is trapped between the ice and the solid rock below. The water has been undisturbed for millions of years, and it's very likely that there are organisms living there.

Prior to studying them, the last thing a scientist wants to do is contaminate the water with bacteria from the surface. So the Russian scientists deliberately stopped short of drilling into the water, which has been named Lake Vostok, while glaciologists, geologists, and planetary scientists from JPL figure out how to build a robot sub that can explore the lake without disturbing it. This is exactly the same challenge that planetary scientists will face on Europa.

Europa represents more than just another place to look for life, though. It serves as a powerful demonstration that the classical concept of habitable zones is not just a shade too narrow, it's absurdly restrictive. If liquid water can exist on Europa, why can't it also exist—not as a moon-wide ocean, perhaps, but at least in pockets—under the surface of Europa's sister moon, Ganymede, which has plenty of ice? Why not on Titan, a moon of Saturn so large that it has a measurable atmosphere? Why not on Triton, a huge moon orbiting Neptune? For that matter, why couldn't the tidal interaction between icy Pluto and its moon, Charon, whose existence wasn't even suspected before 1978, generate liquid water right at the very edges of the solar system?

Such ideas were in the air before the Galileo probe reached Europa. As far back as the early 1990s, the visionary Cornell astrophysicist Thomas Gold proposed that all of these, plus a few asteroids, would be reasonable places to look for life. But before Galileo reached Europa, few of Gold's colleagues took his proposition very seriously. Now, with powerful evidence backing up Gold's theorizing, they're being forced to reconsider.

Life, in short, could conceivably be bursting out all over the solar system. It might be possible to take the Drake equation another, very significant step forward without any need at all for the Planet Finder. Without hard evidence—living organisms from another world or, at the least, fossil evidence that such organisms once existed—that hope is purely theoretical. Hard evidence could come in a decade or so from a mini-submarine on Europa or from a drill boring down into the surface of Mars.

Or maybe the hard evidence is already at hand.

Chapter 12

LIFE ON MARS?

Events like the Japanese attack on Pearl Harbor (for my parents' generation) and the shooting of John Kennedy (for mine) are so emotionally jarring that just about everyone remembers exactly what he or she was doing when the news arrived.

Other events are equally unforgettable, but to a smaller subset of people. I have no memory of where I was when I heard that Elvis had died or when any team won any championship in any sport, but millions of people undoubtedly do. As a science reporter, on the other hand, I can easily recall the context in which I heard that the Hubble telescope's mirror was misshapen or that the Chernobyl reactor had blown up.

I first heard about the marvelous meteorite from Mars one cold, foggy August morning in 1996. I was on an island in the middle of Lake Winnipesaukee in New Hampshire. My extended family takes its vacations on this island, which is owned by the Appalachian Mountain Club, in blissful isolation from newspapers, television, and even the telephone. My father had a portable radio in his cabin, however, and he came to breakfast that morning with a puzzled look on his face. "I just heard something bizarre on the news. They said they'd found evidence of life in a rock from Mars. I must be dreaming, or maybe I'm going crazy. How would they get a rock from Mars?"

I did know that rocks from Mars had been found on Earth—meteorites that were somehow blasted from the Martian surface and lofted across fifty million miles of interstellar space to crash onto our planet. That part didn't sound so surprising. The business about "evidence of life," though, did seem to verge on the crazy. Just six months after Geoff Marcy and Paul Butler had finally moved the Drake equation forward with their discoveries of other planets, was it really possible that someone had already proven that life exists on other worlds?

The previous day Everett Gibson, a geochemist and meteorite expert at NASA's Johnson Space Center in Houston, was on stage at a press conference at NASA headquarters in Washington, D.C., to make a historic announcement. Gibson and the other scientists he worked with had spent more than two painstaking years in the lab using the most powerful tools of geology, microscopy, and analytical chemistry. They'd written up the results of their work soberly and carefully, and submitted them to *Science*, the most reputable research journal in the United States. In the normal course of things, *Science* would have sent the raw manuscript out to three or four scientists with recognized expertise on the topic in question and asked for an independent evaluation. In this case, the implications of the research were so profound that the journal played it safe and called on nine. The reviewers had gone over the paper with an intellectual microscope, questioning every assumption, double-checking every calculation, suggesting changes in wording and tone (always in a conservative direction) that would make the report worthy of *Science*'s implicit seal of approval. Gibson and his colleagues had made the changes. The document was now as responsible a research paper as the scientific community knows how to produce.

In the end, only one sentence—the very last—was at all sensational. Yet that single sentence had brought Gibson, his coworkers, scores of reporters, and dozens of television cameras to this auditorium on August 7, 1996, along with no less a celebrity than NASA administrator Goldin to introduce the scientists to the public. Later in the day, even President Clinton would speak in praise of this research. "Someone told us afterward," recalled an exhausted, exhilarated Gibson when I met him at his office at the Johnson Space Center in Houston a few weeks later, "that in all of U.S. history, only once before has the publication of a scientific paper led to a presidential statement—by Thomas Jefferson, on the reports from the Lewis and Clark expedition. I'm glad I didn't know that at the time."

The final sentence in the *Science* paper had Gibson feeling nervous enough without presidential proclamations raising the ante. Referring to a potato-size, four-pound meteorite that had been picked up on an Antarctic icefield more than ten years earlier, it read: "Although there are alternative explanations . . . we conclude that [the phenomena described throughout the report] are evidence for primitive life on early Mars." The nine scientists whose names appeared on the report claimed, through several converging lines of argument, that they'd discovered nothing less than fossilized Martian bacteria.

After the scientific community digested these startling claims, the consensus was that Gibson and his colleagues had a good argument. Their evidence, laid out carefully and logically, was plausible. But "plausible" isn't the same thing as "convincing," and whether Gibson and his team had met that latter, higher standard was from the outset a matter of fierce dispute.

Critics would challenge several aspects of the meteorite team's research; the team would parry the attacks; the critics would come back with more. A year later there would still be no broad consensus. Long after the *Science* paper appeared, scientists still couldn't agree on whether Gibson and his collaborators had really found what they thought they'd found. And it was unclear when or how the question would ever be settled.

The problem was that the meteorite team had only circumstantial evidence to offer. Depending on a given scientist's personal judgment and taste, that evidence was either compelling or interesting or dubious. Carl Sagan once declared that "extraordinary claims demand extraordinary proof"—that scientific discoveries with a potentially major impact on science or society require more than the usual proof to back them up. Life on another world

presumably qualifies as an extraordinary claim. U.C.L.A. biologist William
Schopf, who appeared at the press conference with Gibson and his partners
as an independent commentator on the paper, even quoted Sagan directly to
argue that it was just that. But, say many scientists, the meteorite team has
yet to present extraordinary proof.

Still, their evidence was reasonably strong. To begin with, virtually nobody disputes the claim that this meteorite came from Mars. Its chemical
composition is significantly different from rocks that formed on Earth; from
rocks that have fallen in from the asteroid belt; and also, as geologists can say
confidently based on samples brought back from the Apollo landings of the
late 1960s and early 1970s, from rocks that formed on the Moon. It does,
however, match eleven other rocks, all of them meteorites, that belong to
what's known as the SNC group. SNC stands for Shergotty (India), Nakhla
(Egypt), and Chassigny (France), where the first three rocks were found.
Their collective origin was somewhat mysterious until the two Viking spacecraft landed on Mars in 1976. The Vikings didn't find life, but they did find
that the gases in Mars's thin atmosphere were subtly different from those on
Earth—there's a higher ratio of nitrogen 15 to nitrogen 14, for one thing.
And they were different in precisely the same way as gases trapped in tiny
bubbles within one of the SNC meteorites, known as EETA 79001. EETA
was clearly from Mars; so, by association, were the others.

The mechanism for getting rocks from Mars to Earth wasn't hard to figure out, either. If you bombard a planet with asteroids and comets, you're inevitably going to send some stones flying. That's especially likely when the
asteroid comes in at a shallow angle, as the laws of chance say one must
every so often, like a golf ball screaming into a sand trap and scattering
grains in front of it.

This must have happened to all of the inner planets. Mars, though, is a
lot smaller than Earth or Venus, and its gravity is weaker. Escape velocity for
a rock (or a rocket) is only 8,500 miles per hour on Mars, compared to
17,000 miles per hour on Earth. (It's even lower on the Moon, and, sure
enough, Moon rocks have fallen to Earth as well.) Chances are pretty good
that bits of Mars would have been flung into orbit more than once during
the planet's 4.5-billion-year lifetime.

Getting them into space is the hard part. Once there, it's inevitable that some of the rocks will eventually wander into the Earth's neighborhood, become trapped by the planet's gravity, and fall. They'll fall pretty much anywhere, mostly in the ocean, since water covers three-quarters of the planet, where they'll never be seen again. Even if they fall on land, they'll most often bury themselves, be covered quickly by plant life, or otherwise vanish from sight.

Not in Antarctica, though. The vast wind-scoured glaciers of west Antarctica are a meteor-hunter's paradise. Rocks that fall here get buried in the ice, but eventually some of that ice, flowing slowly downhill toward the sea, runs into mountains. It breaks like a wave in super-slow motion; the ice piles up and, under the assault of wind that blows up to two hundred miles per hour, erodes away. What's left, like gold nuggets in a prospector's pan, are meteorites, just sitting on the surface waiting to be plucked.

The first Antarctic meteorite was found by chance in 1969, but now that scientists know where to look, they return in force every Antarctic summer. Thousands have been found and returned to the Johnson Space Center's Antarctic Meteorite Center. That's just what happened with ALH84001, the official name of Gibson's rock (the clumsy label means that it was meteorite number 001—the first meteorite of the 1984 season—collected in the Allan Hills ice fields). The collector was a new Antarctic hand named Roberta Score, who noted its striking green color. She realized only later that the gray rock looked green because of her sunglasses. Still, this was definitely a meteorite. Not a Martian one, though: It was identified as a diogenite, a class of meteorites believed to have been chipped at some point in the distant past from the asteroid Vesta.

For nine years that was pretty much all there was to say about ALH84001. Then, in 1993, David Mittlefehldt, a geologist working at Lockheed, which did a lot of government business under contract to NASA, took another look. The meteorite wasn't a diogenite after all. It was a lot more similar, chemically, to the eleven SNC meteorites known to have come from Mars—so similar that Mittlefehldt decided, with concurring opinions from other meteorite experts, that it must be number twelve.

This one was clearly the most interesting, though. The other Martian me-

teorites had been dated—by measuring the decay rates of radioactive elements trapped inside them at the time of formation—at 1.3 billion years old
at the outside. ALH84001 was more like 4.5 billion years old. It had first solidified at about the time Mars itself was born. It lived through essentially all
of the planet's history.

Five hundred million years later, something smashed into the rock—not
sending it into space but riddling it with cracks and melting part of it, a fact
deduced, again, from the decay of radioactive elements. The parts that
melted had their clocks reset at that point. Another four billion years passed.
Life arose on the neighboring Earth, spread, evolved. The dinosaurs came
and went. The common ancestor of both modern apes and humans appeared in Africa. The year (give or take a few dozen centuries) was 16 million B.C. And then another asteroid crashed on Mars, blasting into the
shallow rim of the ancient crater where ALH84001 had survived the earlier
impact. This time ALH84001 was flung out into space. (Nadine Barlow, a
scientist at the University of Central Florida, thinks she's found the site of
that second impact, an oval, shallow-angle crater overlaid on an older one—
or at least she has narrowed it down to two possibilities, both in the planet's
southern highlands.)

Scientists are sure the meteorite spent the next sixteen million years orbiting lazily somewhere between the Earth and Mars; ALH84001's surface
looks to have been cooked with about that time span's worth of cosmic radiation. The cooking finally stopped about thirteen thousand years ago when
ALH84001 slipped inside the protective blanket of Earth's atmosphere.

None of this was all that difficult to figure out—standard meteorite stuff,
really. Things got a lot more interesting, though, when Mittlefehldt and a
Lockheed colleague, Chris Romanek, began looking more closely at the
rock, slicing off paper-thin bits with a diamond-coated saw for microscopic
examination. Embedded in the gray stone they found little blobs of some
brownish impurity. It turned out to be carbonate rock. That was odd. On
Earth, carbonates—limestone, for example, and other rocks made from carbon plus oxygen plus another mineral, such as calcium—form in the presence of water, and the carbon frequently comes from the decaying body
parts of dead organisms. There are other ways of making carbonates, how-

ever, with the carbon coming from carbon dioxide gas dissolved in the water—not an unlikely situation on a planet whose atmosphere was mostly
CO_2. Others don't require water at all: You can get carbon and oxygen and
the other necessary materials directly from other rocks; all you need is an asteroid impact or something like it to melt them all together.

By now Everett Gibson was involved; working with Romanek and other
scientists, he took a careful look at the relative abundances in the rock of
oxygen 16 and oxygen 18—atoms of oxygen with identical properties, but
with the latter containing two extra neutrons in its core. Depending on how
hot the carbonate was when it formed, the ratios of the two types of oxygen
will differ. "We call this the 'oxygen geothermometer,'" Gibson told me
when I visited the paunchy, gray-haired, mustache-wearing scientist in his
office at the Johnson Space Center.

According to what the geothermometer told them, the carbonates had
formed at somewhere between 32 and 175 degrees Fahrenheit—the range
where liquid water could exist. They knew this rock had lived through the
period, between 4 and 3.5 billion years ago, when Viking pictures suggested
liquid water had existed on Mars. Later on, the globules' formation was
dated to about 3.6 billion years. Everything was starting to fall together.

Gibson and David McKay, a senior NASA scientist who had now taken
formal leadership of the Mars meteorite project, began examining the globules more closely, zeroing in on their finest structural details with two types
of electron microscope. McKay was expert at using the scanning electron
microscope, which takes pictures of surfaces. He and Gibson also recruited
another Lockheed employee, Kathie Thomas-Kaperta; she knew her way
around the transmission electron microscope, which probes deep into a
sample. As the name suggests, rather than light waves, both instruments use
electrons that can be more finely tuned to study extraordinarily small details.
Electron beams are like needles, where light waves are like baseball bats.

What they saw, sprinkled liberally around the rims of the globules, were
tiny structures they called "ovoids"—elongated blobs. To any high school
student who has opened a basic biology textbook, their shapes looked un-
cannily like those of bacteria. A more sophisticated observer would realize
that these blobs, at about one hundred billionths of a meter in length, are

about ten times smaller than the smallest bacteria known to science. McKay knew that, and he also knew that formations like this can and do happen through purely mineralogical processes.

Still, he had spent much of his professional career poring over rocks from the Moon and had never seen anything quite like this. So he decided to call in yet another expert. William Schopf of U.C.L.A., who would later appear on the NASA stage with Gibson and McKay, is recognized as the leading authority on fossilized bacteria; despite Steven Mojzsis and Gustav Arrhenius's discovery of bacterial by-products from 3.86 billion years ago, Schopf still holds the record for finding the oldest actual bacterial remains, at 3.5 billion years old.

Schopf went to Houston, took a look, and agreed with pretty much everything McKay and his team had concluded. The blobs sure looked like bacteria, but they also looked like mineral deposits. Anyway, they were pretty small. In Schopf's considered opinion, they weren't worth a damn—unless someone could show they had the internal structure of bacteria or that they incorporated organic chemicals or, preferably, both.

The former would be exceedingly difficult, given the blobs' minuscule sizes, but if it could be done anywhere, the Johnson Space Center, with its powerful electron microscopes, was the place. The latter research would have to be farmed out. Fortunately, Kathie Thomas-Kaperta had a contact at the Stanford laboratory of Richard Zare. Zare was a chemist who had perfected a method for analyzing extremely small bits of matter. He would vaporize a tiny portion of a lab sample with a laser beam—just a few molecules' worth—then measure the chemical properties of those molecules with extraordinary precision.

They sent Zare three bits of ALH84001. He zapped the meteorite fragments. He metaphorically sniffed their vapor. And he detected the faint odor of organic chemistry. Zare had picked up the scent of polycyclic aromatic hydrocarbons, more familiarly known to chemists as PAHs. Nature has plenty of ways of making PAHs, many of which have nothing to do with life. But one of them does: the natural rotting and decay of deceased organisms—plants, animals, bacteria. Half of Schopf's mandate seemed to be fulfilled.

Finally, as she felt her way around the carbonate globules with her trans-mission electron microscope, Thomas-Kaperta made one last discovery. Em-bedded in the rock, like grains of sand frozen in an ice cube, were tiny crystals—some cubical, some teardrop-shaped, some amorphous. Thomas-Kaperta identified them as magnetite and pyrrhotite, the former a type of iron oxide and the latter a type of iron sulfide. There also appeared to be gre-gite, another variety of iron sulfide. All three minerals are naturally mag-netic, and all three are manufactured within the bodies of bacteria on Earth—the B-word again—which use them literally as internal compasses for navigation through the water in search of food.

McKay's team now had four lines of circumstantial evidence suggesting that life had left its mark on ALH84001. There were the carbonate globules themselves, which seemed to have formed in the presence of liquid water. There were bacteria-like shapes. There were organic chemicals in the form of PAHs. And there were magnetic crystals of the same kind known to exist in bacteria.

It was now early 1996, nearly three years since Dave Mittlefehldt had fig-ured out that ALH84001 was Mars meteorite number twelve. It had also been quite a while since McKay and the others decided they had evidence of life. "We actually had the gist of this story in a year to fifteen months after beginning our research," Gibson told me. "But then we spent a year or more trying to refute what we had. Any criticism we were likely to get, we already faced ourselves. We were aware that bacteria have been 'discovered' in me-teorites before. We didn't want to publish until we were absolutely ready."

The truth was that they'd already published a few bits and pieces without coming to any formal conclusions. Thomas-Kaperta had presented the PAH results almost a year earlier, at the 1995 Lunar and Planetary Science Con-ference. She didn't say anything about life on Mars, but her colleagues didn't fail to see the implications. A few eyebrows went up at that point.

Many more would have gone up if she'd gone ahead as planned and pre-sented her new information about magnetic crystals at the 1996 conference. But by now it was time to start writing the big paper—the one that would tie all the evidence together. There wasn't any point to releasing the research piecemeal, and, besides, *Science* wouldn't have any interest in rehashing results that had already been announced. So Thomas-Kaperta, having scheduled her magnetite paper for presentation at the upcoming Lunar and Planetary conference, yanked it from the lineup and pitched in to work on the article for *Science*.

The paper was finished by April and sent in. It went out to the referees, came back for revisions, went back to the magazine. By July it was in its final form, with publication set for mid-August. Two weeks before it was to appear, though, news of the astonishing result leaked somehow. Newspaper and television reporters made it clear that the story was too hot for them to sit on. They were going to go ahead with it, official embargo or not. Unlike *Nature*, though, which had forbidden Michel Mayor to talk to the press on pain of having his 51 Pegasi B paper yanked, *Science* allowed McKay and Gibson to submit to interviews right away. Thus they found themselves, along with Thomas-Kaperta, Zare, Schopf, and Hojatollah Valli, a McGill University geologist who had helped analyze the magnetic grains, under the lights on a NASA stage.

If he'd thought to call Geoff Marcy for a pep talk, what happened next wouldn't have surprised Gibson at all. "It's been overwhelming," he told me later. "We're getting five to six invitations a day for speaking and appearances. We're trying to tell the story the best we can, using the avenues that are best. We've also been to a large number of national and international professional societies as invited speakers—the Geological Society of America, the American Astronomical Society, the university community. We've also been before several major advisory committees, including the Origins committee," he said without stopping for a breath.

Like Geoff Marcy, Gibson and the rest of the team were also having to deal with scientific critics. In Marcy's case, this was a pretty straightforward process. Nobody doubted that the spectral lines in 51 Peg B or 70 Vir B or 47 UMa B—in any of the stars where he'd found planets—were moving. Nobody doubted that this was because the stars themselves were moving. Some theorists doubted that the objects he'd found were planets, although under the conventional definition of what constitutes a planet, even that point was hard to argue. And even if some of them were brown dwarfs, it was hard to claim, as David Black tried to do, that they all were.

The Mars meteorite team had a much tougher job, though, as they candidly admitted in their paper. The reason, they acknowledged, was that none of their evidence was close to being definitive. The last two sentences of their paper read:

> None of these observations is in itself conclusive for the
> existence of past life. Although there are alternative ex
> planations for each of these phenomena taken individu
> ally, *when they are considered collectively, particularly in*
> *view of their spatial association,* we conclude that they
> are evidence for primitive life on Mars. [italics mine]

In other words, you don't necessarily have to be convinced that life is the only or even the best explanation for each of the lines of evidence taken alone—PAHs, ovoids, and magnetic particles, all jammed close together within carbonate globules. You only have to agree that it's plausible for all of them. And then you have to decide that four "maybes" add up to a "probably."

As Gibson, McKay, and the others found out pretty quickly, many scientists don't buy that equation: "[T]he conclusions of the NASA group are hypothetical," Thomas Ahrens, a planetary scientist at Caltech, told *The New York Times.* "[T]he interpretation of this Allan Hills meteorite, in a way, is a house of cards that could collapse if any of the inferences are wrong."

Plenty of scientists felt that some of the inferences were wrong and weren't shy about saying so. "We knew when we made this proposal that it was going to receive considerable criticism," Gibson told me, a bit wearily. "Any great idea in science takes its heat before it's accepted. So we're continuing to spend a lot of our time in research and also in responding to critics." Gibson pointedly failed to note, of course, that incorrect ideas get as much heat as great ones.

Not a single one of the meteorite group's claims escaped strong challenge—and no challenge failed to elicit a sharp counterargument from Gibson and his group. Take the PAHs. They are indeed produced by rotting animal matter—but PAHs have also been detected in interstellar dust grains and have been found in meteorites from the asteroid belt. "In none of those cases have they ever been interpreted as being biological," Schopf said at the press conference. "Assuming that they're not contamination from industrial pollution on this planet, the first guess would be that they're probably non-biological, just like PAHs that occur in other meteorites. The burden of proof is on those who claim that they're biological."

Jeffrey Bada, in the NSCORT group in San Diego, agreed. "Rather than being biology," he said, "I think it's more likely that they're either abiotic compounds derived from stuff like dust or comets falling on Mars, or that it's terrestrial pollution." A postdoctoral student of Bada's named Luann Becker analyzed the PAHs found in Antarctic ice and determined that many were chemically similar to the ones in ALH84001 and to those in EETA79001, the first Antarctic Martian meteorite (the name means it was the first meteorite picked up in 1979, at the Elephant Moraine). Maybe meltwater percolated in through cracks in the rock sometime during its thirteen-thousand-year nap in Antarctica.

No way, said Gibson. The PAHs are found in higher and higher concentrations toward the center of the meteorite. That's just the opposite of what you'd find if they soaked in from outside.

Not true, said Becker. The outer surface of the meteorite, baked hard by air friction during its passage through the Earth's atmosphere, isn't as absorbent as the porous rock inside. It couldn't soak up as much PAH-laden water.

Bad reasoning, countered Gibson. The meteorite shows no physical signs of weathering, as it should if water had passed through it.

And so on. Each side kept landing blows, but nobody could score a knockout. You couldn't prove the rock had been contaminated on Earth. You couldn't prove it hadn't.

The same applied to each of the other lines of evidence. The magnetic crystals looked very similar to the ones produced by bacteria, said some scientists. Not *that* similar, insisted others. Bacteria a hundred times smaller than any known on Earth were a highly dubious proposition, said Schopf. At a mere one hundred billionths of a meter long, they might be too small to process energy—too small to live. Not at all, said Gibson. "We got hold of samples from deep drill holes in the Columbia River basalts [one of the places where biologists first discovered deep subsurface bacteria] and looked with our SEM, and we began to see the same shapes and sizes we'd seen in our Mars rock—but this is from a terrestrial system that is only just now being characterized. The experts had said you couldn't have living things this small. You couldn't pack in enough of the right chemistry—amino acids and things. But these are living organisms, and they are that small. They exist."

Perhaps, but until there's a report in a refereed scientific journal, they don't exist officially. And even if they do, argue the critics, until you see cell walls or real evidence that the ovoids are bacteria, it's more reasonable to assume they're purely mineral in origin. A judgment call once more.

Similarly for the carbonates themselves. Gibson and company had good chemistry-based arguments to show that they were formed at liquid-water temperatures. That hardly proved that they had any relation to life; carbonates can form out of mineral-rich, lifeless seawater. Beyond that, a paper appeared in the British journal *Nature* a month before Gibson and friends' report came out: It used good chemistry-based arguments favoring a formation temperature for the carbonates at about 1200 degrees Fahrenheit, much too hot for life to survive.

Gibson had not one but four different counterarguments for that one. First, the PAHs would have been destroyed in such an oven (though if they weren't there to start with, this wouldn't apply). Second, the iron sulfide particles would have degraded at such temperatures. Third, the carbonates are of several different types; high-temperature formation would have produced only one. Fourth, since the press conference, a British team had discovered methane in the meteorite—more evidence for life, incidentally, and yet another substance that would have been degraded by high temperatures. "Now, you may find a high-temperature mineral somewhere in the system," he admitted. "But who says that hasn't been carried in by the fluid that deposited the minerals from which the carbonate was formed?" Good point. Another judgment call, which could be made either way depending on what really amounted to nothing more than personal scientific taste.

By the summer of 1997, nearly a year after the *Science* paper appeared, nobody had yielded any ground. Jeffrey Bada, William Schopf, and the other skeptics were inclined to disbelieve in Martian microbes—something they considered an extraordinary claim—unless forced to do so by extraordinary evidence. Gibson, McKay, and the rest thought they'd started out with a strong case and hadn't been persuaded otherwise. "We're still on solid ground," Gibson told me. "Data are coming in that are beginning to support what we have. There are vocal critics out there who don't want to accept a far-fetched idea like this one—and that's their right. But we've debated them at national meetings, and we haven't run into anything we cannot handle. That's not boasting—just the facts as we see them. Anyway," he said mysteriously, "we're sitting on some data we're pleased with, that support us further."

Unless it's cell structure, though, or unambiguously biology-related chemicals or some comparable smoking gun—unless the evidence rises above the plausibility threshold to that of probability—the critics will undoubtedly remain unconvinced. Jeff Bada, for one, not only wants to see biological chemicals but also wants to see them with a "handedness"—a chemical asymmetry built into all organic chemicals—that's the opposite of what's seen on Earth. (All of ours are of the type called left-handed because when you send a light beam through a solution of them, it twists toward the left.) Only then will he be convinced that the organics in ALH84001 truly came from beyond the Earth.

That's probably too harsh a standard, though. Life on Mars could have ended up left-handed as well; it had a fifty-fifty chance. Recent observations of interstellar dust suggest that the odds might be better than that: For reasons nobody understands, Nature's preference for left-handed molecules seems already to be evident in deep space. "The truth," said Bada, "is that we can argue about what these meteorites tell us until the cows come home, and we're not going to ever, in my opinion, have the conclusive evidence one way or another. That means we really do have to go there."

As he spoke, in December 1996, two spacecraft were already on the way; one would be going into orbit to map the surface in great detail, in part to look for likely future landing sites, and the other would drop down to put a two-foot-long, six-wheeled rover named Sojourner on the surface to nose around and study the rocks and soil.

Sojourner wouldn't be looking for evidence of life; unless Viking-project scientists had completely misunderstood the data sent back by their twin

landers in the mid-1970s, the Martian surface has no life. No one even bothered to outfit Sojourner with the sensors that could detect it. The little robot
cart's job was more basic than that: tool around the landing area, roll up to
some interesting-looking rocks, bombard them with X rays, and see what sort
of light bounced back. The reflections would tell geologists what chemical
elements were present in the rocks, and, along with the pictures, would perhaps give the scientists some clues to their geologic history.

After traveling for eight months across interplanetary space, screaming
into Mars's thin atmosphere at seventeen thousand miles per hour, and
bouncing to a stop on the red planet's surface, that's exactly what Sojourner
did. The little car exited its parent craft, Pathfinder, and cruised across the
red, dusty soil at a stately two feet per hour to examine rocks named, in an
unusual burst of silliness among Jet Propulsion Lab scientists, Barnacle Bill,
Scooby Doo, and Yogi.

The NASA scientists had guessed that the landing site, an ancient river
delta called Ares Vallis, would prove interesting. Water had flowed there—at
least that's what orbital photos suggested. Sojourner confirmed the water but
told them that it had not just flowed but roared. The rocks littering the landscape were very different from one another and had clearly been carried
there by powerful currents.

Much more tantalizing for the prospect of finding Martian life someday
was the cameras' discovery of cracks in the planet's surface and what look
like salt deposits on the rocks, evidence that water not only moved across
Mars but stood for a period of time. Virtually all the scientific scenarios for
the origin of life call not just for liquid water but—a point assumed but
rarely made explicitly—for water that stays in one place for a while.

Sojourner can't do much more than take pictures and sniff rocks, though.
Ideally, scientists want to get their hands on Mars rocks that haven't been lying in Antarctic ice for thousands of years and study them in detail. JPL is already working on a mission like this—fully automated, like Sojourner, since
the cost-conscious Dan Goldin has insisted, and reiterated after presenting
ALH84001 to the world, that the agency won't be sending humans to Mars
anytime soon. A sample-return mission could happen by the year 2005 or
even sooner if NASA can get its hands on the funding.

These samples will still be scratched from the Martian surface, though, where life almost certainly hasn't been for billions of years. If it still thrives, Martian life is deep underground. But Jeff Bada, who has been working with a team at the Jet Propulsion Lab to design a (probably quixotic) amino-acid detector that could go on an intermediate mission, scheduled tentatively for launch in 2001, has an idea how to get around that. "Drilling will be pretty difficult," he said. "You need heavy-duty equipment. My feeling is we need to blast." Bada wants to send up a space probe that carries a little cruise missile. "We need to make our own little crater and then rush in and sample it as soon as we can."

And then, if Everett Gibson, David McKay, or their scientific descendants find incontrovertible evidence of Martian life, past or present, one question will remain: Did life arise independently on both worlds? If Martian microbes are chemically distinctive from those on Earth, it will be clear that they did. But if Mars bugs seem to be close cousins to those on Earth, Drake-equation scientists will have to consider another possibility. If an asteroid could blast a rock from Mars to Earth sixteen million years ago, it could just as easily have done so four billion years earlier, at a time when a rock like ALH84001 could have contained not fossilized bacteria but living ones. If rocks can fly freely from Mars to Earth, then why must we assume that life began on Earth at all? It's not inconceivable that every ounce of life on the planet—plants, animals, bacteria, fungi—is the result of a contamination accident, an infestation of microbes that leaked from a meteorite onto our then-pristine planet and completely took it over. We may all be Martians.

That would deal a fatal, final blow to terrestro-centrism. The proposition that Earth is the only place in the neighborhood where life began is already under serious assault. Now we have to wonder whether our planet didn't even have the stuff to get life going or whether the original biota of the Earth was outfought by an invading army of Martian microbes.

Jarring as this would be to humanity's view of itself, it would be nothing compared to the discovery of intelligent life elsewhere in the universe. New planets and alien bacteria are important steps on the road to unraveling the

Drake equation. But the ultimate solution is to find extraterrestrials that think, that have developed technology, that use radio waves for communication. That's what Frank Drake had in mind when he pointed the eighty-five-foot radio telescope at the West Virginia skies in 1960 to launch Project Ozma, and that's the goal his colleagues are still pursuing.

Chapter 13

LISTENING FOR EXTRATERRESTRIALS

On Halloween night, 1996, I checked my E-mail and found the following message:

> Observing finally began late this afternoon. Observations of 51 Peg at S-band (that is, the upper range of our 1.2–3.0 GHz search space) were carried out before dinner. The aliens remain elusive. But 47 U.Ma. is on the schedule for tonight, I believe. By morning we may know of their plans for our planet and our women.
> Cheers, Seth

Talking to the author of this message, Seth Shostak, is a bit like talking to Robin Williams—if Williams had a Ph.D. in astrophysics from Caltech. Shostak can't get through more than a couple of sentences without throwing in a comic aside or two. A reference to bad 1950s science fiction movies is not unusual. His looks match his mind-set: Shostak is trim, with graying hair that falls in bangs on his forehead and a perpetual expression of repressed mischief, suggesting that he's thinking of a one-liner even if he's not going to say it.

But Shostak is also a serious scientist, a highly trained radio astronomer who made pioneering observations of distant galaxies back in the late 1960s and went on to join the faculty of the University of Leiden in Holland. Now he's professionally engaged in looking for aliens. Ever since Frank Drake gingerly broached the subject in 1960, that field has been accepted as legitimate by a wide range of scientists. Philip Morrison, Carl Sagan, Charles Townes (the inventor of the laser), Freeman Dyson, Ronald Bracewell, George Gatewood, and dozens of other respected researchers have been involved, at some point or other during their careers, in SETI and the Drake equation.

Still, it helps to have a sense of humor about the whole thing. After four decades of searches involving dozens of astronomers putting in thousands of hours of observing time on scores of radio telescopes, SETI has not come up with a single confirmed signal from an alien intelligence. Beyond that, it's impossible to predict when one might appear.

It isn't just that most of the Drake equation is still essentially pure speculation, although that's certainly true. Even assuming that the Milky Way is home to ten thousand civilizations capable of communicating with us—a number Frank Drake considers plausible—it could take centuries to find just one of them. The sky has millions of candidate stars, and you won't detect a thing if you're looking in the wrong place.

Even if you're looking in the right place, the radio spectrum has millions of channels. Tuning to the incorrect frequency is like turning on MTV and expecting to see *Sixty Minutes*—and in this case there's no *TV Guide* to tell you what the right frequency is. And then, if you have the right star and have managed by incredible luck to tune into the right channel, you might have the wrong time of day. If the aliens are on the air for only one hour out of every ten and you listen for only an hour, you'll almost certainly miss the broadcast. Given this piling on of uncertainties, the signal might be here already, and we still might not pick it up for a thousand years.

During my three-day visit to Green Bank, West Virginia, in late October 1996, the odds of detecting an alien broadcast were not merely small, they were zero. I had gone to the National Radio Astronomy Observatory—the same installation where Frank Drake had launched Project Ozma thirty-seven years earlier and where he had scrambled to find an organizing principle for the first SETI conference—for the inauguration of a new search for extraterrestrial intelligence. It was part of a larger overall effort known, for reasons that will become clear, as Project Phoenix. (In its previous phase, Phoenix had had Shostak pacing the floor all night, waiting for a telephone call from Australia that would confirm or deny that they'd found E.T.)

Unfortunately, the 140-foot-diameter radio telescope the Phoenix team was using had been aimed since early that morning not at the sky but at the horizon. That's the position it assumes when NRAO engineers are installing or removing receivers, the radio telescope equivalent of the light detectors on optical telescopes. Radio astronomers study all sorts of phenomena—wispy filaments of electrically charged particles whipping around the core of the Milky Way; blobs of cool gas hovering between the stars, waiting patiently to condense into stars themselves; particle jets shooting from the violent hearts of quasars at nearly the speed of light; bolts of lightning sizzling through the noxious stratosphere of Jupiter; alien radio programs. Each requires a different sort of listening device. You can't pick up television on an AM radio; you can't tune in cellular phone calls on a TV; you can't get CB radio on a cellular phone. It's the same in radio astronomy; you need to switch over to a new receiver for each new project. Getting the receiver to work smoothly with the rest of the system is something like hooking up stereo components for the first time or setting up a new computer yourself. Sometimes it just plain doesn't work, and you can't for the life of you figure out why. As one of the engineers explained it to me, "We've been trying all morning to get the computers to send instructions to the receiver. It just answers *phhhhht.*"

Long after dark it was saying the same thing. Still, there was plenty of work for the Phoenix team to do—testing its own computer equipment, reviewing data tapes from two weeks earlier when engineers had attached the detector for a test run (it had worked back then, without a hitch), and generally just checking things out. Much of the work was going on at the 140-foot, the largest of several radio antennas that line NRAO's main traffic artery—a two-lane roadway that passes through a landscape of meadows and woods so bucolic that it could be part of a national park except for the enormous white dishes planted in the fields alongside.

Ordinary cars aren't allowed here; their spark plugs give off static that can destroy a sensitive radio observation. So the astronomers drive observatory diesels, ride observatory bikes, or walk—the last two means of transportation somewhat hazardous at night since the road is not illuminated.

When it was built back in the early 1960s, this was one of the largest, most sensitive radio telescopes in the world. As I pulled up to the concrete building that serves as the telescope's massive base and also houses its control room and computers, it seemed plenty big enough still.

From 1971 to 1991, though, the 140-foot had been dwarfed by a 300-foot dish a quarter-mile or so down the road. The 300-foot had collapsed suddenly one night, the victim of metal fatigue, but it would soon be replaced by an even bigger 330-foot dish to be called the Green Bank Telescope. The GBT's mount was nearing completion when I arrived. It was already taller than the 140-foot even though its dish was still lying in prefabricated pieces on the ground. Both radio dishes are dwarfed by the 1,000-foot dish built, permanently immobile, into a bowl-shaped valley at Arecibo, Puerto Rico.

The 140-foot is therefore a second-class instrument. Not much cutting-edge science gets done here. That's just fine with the Search for Extra-Terrestrial Intelligence folks. They'd love a bigger, more sensitive telescope, but while their colleagues generally agree that SETI is a worthwhile project, they also agree that other projects should come first. The big dishes are crucial in helping to answer most of the major mysteries of astronomy: How big is the universe? How old is it? What is it made of? What are the quasars? How were the galaxies born? What is the large-scale structure of the cosmos? Unlike SETI, these other questions have a reasonable shot at being answered, at least in part, sometime in the next few decades. And so in the constant competition for telescope time, SETI is generally put at the end of the priority list.

An old dinosaur like the 140-foot isn't really all that much in demand anymore, though. As a result, it's easier to get time here than it is on some of the newer, more powerful instruments. And given the low probability that a signal will come in on any given day, time is what Project Phoenix desperately needs. This sojourn in Green Bank is considered a major coup. Phoenix had been given a week now, in October, then two weeks in December. In January they'd get fully half of the 140-foot's time, and by the end of the year the Phoenix crew would have it all to themselves for several months. When another, bigger dish became available, as the giant Arecibo telescope would for a time in 1997 and as a radio telescope in France would in 1998, they'd simply pick up their equipment, conveniently installed in a heavy-duty steel trailer that was located about thirty yards to the left of the 140-foot but that could be made ready for shipping on short notice.

Seth Shostak and his colleagues on Project Phoenix are scientific vagabonds, but they do have a permanent home base—not a university or a NASA lab but a private institution called the SETI Institute. The grandly named institute occupies a couple of suites in a suburban office park in Mountain View, California, alongside dentists, graphic arts firms, and other small businesses. "When people come here, they expect to see people with headphones on, listening for E.T.s," Shostak told me when he showed me around headquarters in September 1996, about a month before I caught up with him in Green Bank. "They're usually kind of disappointed."

The nonprofit institute was founded by Frank Drake in 1984 as an administrative device to streamline research in all aspects of his equation. Government grants have to be administered by someone—a university, for example, or even the granting agency. The administrator rakes off some money, but a small nonprofit organization needs to rake off less. While the SETI Institute manages money for scientists all over the world, it runs only Project Phoenix directly.

Phoenix started out as a NASA-funded program called the Microwave Observing Project (the radio waves it looks at are the shortest radio waves, which happen also to be ideal for heating up food). SETI had long since earned the respect of scientists, including the blessings of the National Academy of Sciences, as well as the interest of the general public, so this was a natural project for the space agency to underwrite.

Unfortunately, SETI was also identified, to the despair of its practitioners, with UFOs and their associated themes of crashes in Roswell, New Mexico, government cover-ups, alien bodies on ice in Air Force hangars, crop circles, and the like. It's very easy to confuse the science with the silliness. It's especially tempting to do so when political posturing and taxpayers' money are involved. Thus, in 1978, after two decades of NASA involvement with SETI, Senator William Proxmire decided to lampoon the effort with his infamous Golden Fleece award.

The point of the award was to make fun of projects that Proxmire thought were a waste of money (they frequently involved sex research). A few years after bestowing his award, he went a step beyond ridicule. He tacked onto an appropriations bill an amendment that forbade NASA from funding SETI research. The bill passed. SETI was cast adrift. It was only after Carl Sagan went to Washington to explain to Proxmire personally the scientific rationale behind the search that the senator relented and allowed funding to resume.

But in 1993 it happened again. This time it was Nevada senator Richard Bryan who ranted against government funding of SETI. "Not a single Martian has yet been found," he thundered, and he demanded that NASA cease and desist its support of this foolish project. Bizarre words coming from a senator whose state has declared one of its roads the "Extraterrestrial Highway" to celebrate all the UFO sightings made along the roadway.

"He was trying to make a name for himself as a budget cutter," Drake told me on my visit to the institute, not bothering to hide the exasperation in his voice, "but every time he came up with an item to slash, it involved jobs in someone else's state. He didn't want to make enemies; SETI didn't involve any jobs." NASA, which was having enough trouble getting money out of Congress for its higher-profile programs, didn't argue. When someone asked Daniel Goldin about SETI after his Origins speech to the American Astronomical Society, his answer artfully straddled a rhetorical fence. "It's a fine program," he said, "but we feel that it's best left to the private sector."

In fact, the SETI Institute was already tapping into private money before Bryan pulled the government plug; the cutoff robbed the institute of only half its grants. Now it's back to full funding from donors that include Paul Allen, who cofounded Microsoft with Bill Gates. "We're actually more secure than we were under NASA," Drake says, "since their budgets have to be approved from scratch every year." The Microwave Observing Program (MOP) arose from its own ashes and was renamed Phoenix.

From the start, Phoenix was designed to combine the best features of all previous SETI searches. Some, like the BETA survey out of Harvard University, scan the entire sky on the theory that you can't predict where a signal might come from. Others, like the University of California, Berkeley's Serendip survey, are aimed at specific targets because lingering on a particular star makes it more likely that you'll pick up a fainter signal.

Phoenix—or MOP—was going to do both: an all-sky survey operating from the Jet Propulsion Lab and a targeted search aimed at the eight hundred or so nearby Sun-like stars. The end of NASA funding eliminated the JPL half of the project entirely. Phoenix is a targeted search only.

That's probably just as well. While nobody can say anything for sure about what an alien signal will look like, it will almost certainly be vanishingly faint. "With the 140-foot telescope," Shostak told me, "we can see really weak signals. We can pick up a cellular phone operating on Pluto. But that's probably not good enough."

Just how strong a signal will be depends in large part on whether it's aimed in a tight beam directly at the Earth or is just random broadcasts leaking out into space. SETI enthusiasts often trot out the statistic that Earth's earliest TV broadcasts, blasting off into space in all directions at the speed of light for the past fifty years or so, have by now spread over a sphere fifty light-years, or nearly three hundred trillion miles, in radius. *I Love Lucy* has reached about fifteen hundred stars so far and reaches another hundred every year.

The problem, as Frank Drake explained during my visit to the institute, is that these broadcasts weaken rapidly as they travel. "You could pick up only one of our TV broadcasts with the kind of radio telescopes we use at a distance of about four light-years," he said. That's the distance to the Alpha Centauri system—the closest to Earth.

Moreover, television signals consist of two parts, a very narrow, powerful carrier signal and a much broader, weaker signal that carries the picture information. It's the carrier signal you can see at four light-years, not the information signal. If the aliens are pretty much like us, they'd better be on Alpha Centauri—and if they are, the only part of their leaky broadcasts SETI scientists could detect would be the carrier signal. They'd then have to go out and commandeer much bigger telescopes to get any sort of information.

Perhaps the aliens are a little more active about interstellar relations than we are—an assumption as good, in the absence of any information to the contrary, as any other. In that case, they might send out deliberate beacons, focusing lots of energy into a tight, efficient beam, something like radar, to try to attract the attention of other intelligences. If they use radio waves to do it, then the number of stars audible to the radio ear of the 140-foot jumps from one to one thousand.

This number involves leaps of faith on several levels, however. Who's to say, first of all, that aliens want to attract our attention? The physicist Freeman Dyson, of the Institute for Advanced Study in Princeton, has suggested that their primary interest will more likely be their own survival. To feed ever-increasing energy demands, he suspects, aliens will eventually destroy extraneous planets and use the debris to build shells around their stars, to trap and tap every bit of stellar power. If so, what were once visible stars will be replaced with "Dyson spheres," immense spheres of rock glowing dully in infrared light from the small amount of heat that would leak from them. Astronomers have actually searched for Dyson spheres. So far, no luck.

It's also presumptuous to think aliens would broadcast on just the radio frequencies we happen to be listening to—or use radio at all, for that matter. Here on Earth we're already switching from broadcast signals to fiber optics and cable for much of our information transmission. Who knows whether we'll even use radio a century from now, even for interplanetary communication. Maybe lasers will turn out to be better. Maybe the aliens know this already and are flashing lights at us instead of radio waves—or maybe E.T. is using radio but is choosing wavelengths different from the ones we're looking at.

The SETI researchers realize this, but like those who are looking for evidence of more primitive life on other worlds, they can only do so much. Their first assumption has to be that aliens' communications strategies, like their biochemistry, will be pretty much like ours. If this assumption is wrong and listeners find nothing after a good, thorough search, then they'll consider moving on to a different line of reasoning.

The Phoenix team has made some basic assumptions about how aliens might behave and have adapted the equipment they have to take maximum advantage of it. When the telescope was finally up and running, it would point to the first candidate star and sit there, listening, for about 550 seconds. Radio signals would begin pouring in, whether or not E.T. was on the air — for while stars are much quieter in the radio band than they are in the optical, they and the interstellar space between us and them crackle with some radio noise.

When it's working, the curved 140-foot dish funnels these radio waves into the receiver, which converts them into a set of frequencies that its electronics can read, understand, and translate into a digitized version of the original. Digitizing the signal is crucial. The receiver is listening to a big chunk of the radio spectrum — it's tuned into millions of channels all at once. An alien broadcast, though, will be coming in on only one of them, most likely. So rather than listen to them one by one, the Phoenix team analyzes them all at once. That's a task only a high-speed computer can handle; thus, the signal has to be turned into a digital form that a computer can understand.

So the signal is digitized, split in half, and piped out to the trailer, into a specialized sort of computer called a multichannel spectrum analyzer. Inside this device, each half-signal is split again into one hundred channels, each spanning 74 megahertz on the radio dial. It's then split again, into a total of about fifteen thousand channels, and each of these is split again into one thousand.

The trailer's electronics are now humming with two different parallel streams of data, each split into approximately fifteen million separate radio channels. Finally, the analyzer looks at each and every channel through two different filters, for a total of almost sixty million individual radio channels, any one of which might be carrying a message from another world.

Any one of the sixty million channels could also be carrying a signal from just about anything else — a plane, a satellite, a radar beacon. By the time observations have begun, though, the astronomers have already listened extensively over several weeks to the local noise. They've built a catalog of signals they know are not E.T. and fed them into yet another special-purpose computer called the Follow-Up Detection Device. "It's the most powerful computer in, well, West Virginia," said Shostak. "At least that's been suggested. I have no idea whether it's true." The FUDD is powerful enough, in any case, to scan through all fifty-six million channels, pick out any regular, repeating signal (static won't do), and compare it to the catalog. Familiar signals are automatically trashed.

But if a signal looks both promising and unfamiliar, the FUDD alerts a second telescope to swing into action. It's this second telescope that makes Phoenix much harder to fool than any SETI search in history. The second telescope is located in Woodbury, Georgia, seven hundred miles to the south. The distance means that if the signal is due to local interference that hasn't yet been catalogued, it won't even register at the backup dish.

This weeding-out process is an efficient way of getting rid of spurious signals. During their 1994–95 run at the 210-foot Parkes radio telescope in Australia, the Phoenix team made a total of 24,074 observations of 209 target stars. They detected 148,949 radio signals, of which 115,570 were rejected out of hand as known local interference. Of the remainder, 18,255 were sent out to the second telescope, which confirmed only 39 of them. And of those 39, every one turned out to be a satellite. Not a real alien signal in the bunch.

The crucial, final test for eliminating that last thirty-nine is an especially ingenious one. If a source is truly celestial, its radio frequency is going to be affected by the rotation of the Earth. If it's just on the horizon, for example, that means Earth is rotating toward it, and its radio waves should be blue-shifted, just as 51 Peg's light is when it's wobbling toward us. If the source is overhead, it shouldn't be shifting at all. But since the second telescope is at a different place on the Earth's surface, the direction of the source, and thus the wavelength shift, should be different, too, in a predictable way.

At each telescope, moreover, this wavelength shift should change as the angle between telescope and signal source changes. The change shows up as a line that moves diagonally across the computer screen, as the red- or blue-shifted signal moves steadily from one wavelength to another. If the source isn't shifting in the right ways at both telescopes, it isn't celestial.

It turns out that there should also be a Doppler shift caused by the rotation of the planet where the signal is coming from. "You could in principle gauge the rotation rate of the planet that way," Drake told me. If there were many beacons on the planet, you could learn even more—where the oceans were, presumably, because they wouldn't have transmitters; how big the planet was, because beacons at high latitudes would be in view for shorter periods of time; and so on.

"Because they didn't use this Doppler drift," Drake said, "I don't consider any detection over the past thirty years very serious, including the WOW signal." He was referring to a mysterious blip of radio noise that was snagged by a SETI search based at Ohio State University. No one was watching one night in 1977 when the signal zipped in from space, a tight blip of radio energy that turned on, then off in a way that suggested it might be artificial. An astronomer named Jerry Ehman saw the blip recorded in a computer printout the next day and wrote "Wow!" in the margin.

Unfortunately, it never repeated—and without a Doppler drift, there was no way of knowing whether or not it was coming from the stars. "The only signals I take at all seriously," said Drake, "are thirty-seven or so recorded by the Harvard search, which do measure Doppler drift." These signals were never repeated, either, though. They remain possible but not confirmed evidence of intelligent E.T.s. To date, there has never yet been a confirmed artificial signal from the stars.

Fortunately, there's a way to test how well this so-called Doppler drift test works without actually nabbing E.T. "Here, take a look at this," said Jill Tarter, pointing at the computer screen. I had finally made it inside the control room of the 140-foot telescope at Green Bank to find several of the Phoenix scientists, including Shostak, seated at their computer terminals. Tarter, a tall, short-haired, serious woman, headed the MOP project for NASA before Congress pulled the plug, and she's now Project Phoenix's chief scientist, and a neighbor and close friend of Geoff Marcy's. She's also the scientist model for Jodie Foster's character in the movie *Contact*. Tarter had arrived at Green Bank directly from Rome, where she'd taken part in a Vatican symposium on Life in the Universe. (It was at this conference that the Pope had, to everyone's surprise, rescinded the Church's long-standing objections to Darwin's theory of evolution.)

Despite a bad case of jet lag, Tarter insisted on being here at the telescope, ready to start observing the moment the technicians got the thing fixed. Suddenly, she punched some keys on the computer, and an image appeared on the display. Most of the screen was black, with a few random dots of white—radio static. But there was also a distinct white line stretching from the top of the screen to the bottom, angling slightly from left to right. It was a textbook example of Doppler drift.

"That's Pioneer 10," she said. The space probe, launched in 1972, was now well beyond the orbit of Pluto. At that distance it's far enough away so that Doppler drift is indistinguishable from that of a distant star. For a moment I thought the system was finally working—but this had been recorded a couple of weeks earlier, during engineering tests. "Once we're up and running," said Tarter, "we'll look at Pioneer 10 once a day, just to calibrate the equipment."

This brought to mind a question Shostak had been meaning to ask: "What star do you think we should look at first?"

"47 UMa," answered Tarter.

"I don't think it's on our list," Shostak said.

"It can get on the list. We can't avoid looking at the systems where we *know* there's a planet," she replied. "That would be ridiculous. And 47 UMa smells most like our solar system."

With any luck the telescope would be working by the next morning. "We'll be okay," said John Dreher dryly, "as long as we don't show any fear. It can smell fear." Up until now, Dreher, a slightly overweight man in his forties, had stayed silent—not a typical state for him, I would learn as the night wore on. Like Shostak, Tarter, and Drake, Dreher started out as a conventional radio astronomer. He was on the faculty of M.I.T. for a while, but for a number of reasons he didn't fit in. One, almost certainly, was his personality, which one could easily see would ruffle academic feathers. Dreher doesn't suffer fools easily, nor does he keep secret his opinions about who is and who isn't a fool. Moreover, he is prepared to issue opinions on almost any conceivable topic with absolute confidence. "Don't you hate people like that?" Shostak would ask later. "Especially when they're almost always right?"

It was clear, watching these two in action, that over years of working together they'd settled into the kind of relationship you see in a long-married couple—mutually respectful but also capable of a little bickering at times. As Frank Drake had realized early on, SETI is a peculiar sort of science to work on. Butler and Marcy searched for otherworldly planets for eight years without success; in SETI you might work for an entire lifetime and have absolutely nothing to show for it. It takes a peculiar personality to work on SETI for long, and these two, Shostak and Dreher—a natural comedian and a natural curmudgeon—certainly qualified.

As they sat at their keyboards, Shostak and Dreher talked first about technical issues, but then Dreher drifted into other topics. The conversation stayed on the subject of SETI, but it was red-shifted to a more philosophical level than the nuts and bolts of how the search actually works.

"What I really find tough to project," said Shostak, "is the way we'll feel if we really find something. It will turn everything upside down, don't you think?" Evidently his experience walking the floor all night waiting for word from Australia the year before had unsettled him.

"No, I don't," said John Dreher. "I think it would take one or two hundred years for a detection to really sink in. I mean, we talk about the great Copernican Revolution, but what did it really mean to the average guy when we found that the Earth really goes around the Sun instead of the other way around?"

"I guess you're right," said Shostak. "It was a paradigm shift but at a very high level. It'll be like the Martian meteorite—big news for about a week. Okay, then how long will it take you before you're confident about a detection?" It was the old Peter van de Kamp question again.

"A week," answered Dreher without hesitation. "But I would still wake up at night worrying about whether it was a real detection or a bug in the system."

"Or a hoax," added Shostak. In fact, it's hard to imagine any sort of hoax that could fool the Phoenix crew. The only way to duplicate a signal coming from space would be to go out there and send it. Even a signal bounced off the Moon—an undertaking beyond the scope of a casual hoaxer in any case—would be easy to spot. Hackers are a bit more worrisome. They got into the computers at Lick Observatory in 1995. (The SETI people know this because in addition to his work at the Institute, Frank Drake is a professor emeritus of astronomy at the University of California, Santa Cruz, where Lick has its administrative headquarters.) "It was a real mess," Drake said. "It took months to clear everything up." At the moment, the Phoenix system isn't connected to the Internet. But eventually the scientists want to be able to observe remotely, so they'll have to construct an electronic fire wall.

In the control room of the 140-foot, Shostak was still trying to get John Dreher to take his philosophical musings seriously. "Okay, John, fine, we agree an E.T. detection won't matter to the average person. You're not an average person. How would *you* feel."

Dreher rolled his eyes. "I'll be a happy guy, okay?"

"Yes, but won't it tear apart your worldview?"

"No, because I already think the universe is teeming with civilizations. We're here. Why not a zillion other civilizations? Yes, knowing is different from believing, so I promise I'll smile when it happens. But I won't be surprised. Frank Drake's a great scientist, but I think these numbers he pulls out of his hat, like ten thousand civilizations, are an affront to common sense. Why shouldn't there be two hundred million?"

"I can go through the numbers, John. You know the numbers."

Later, Shostak would explain what he'd been trying to get at. "He kept giving me intellectual answers. It's not an intellectual thing. The question is, if you've been doing an experiment for ten years with no result and suddenly you have one, how do you feel? You ask people, and invariably they lapse into historical analogies or talk about what procedures they'd follow. I really think it would be a huge, emotional monkey wrench. I don't think I'd be able to get over it. The average person can escape into everyday life. For us there will be no escape. How we'll deal with it is something SETI people never really talk about. Everything else would turn into this veneer of irrelevancy, and under it would be this incredible fact."

At first the fact of a detection would probably be all the scientists would have. Even if they picked up more than just a carrier signal, the message would likely be incomprehensible. "It would be poetic justice if we were the ones to decode it," John Dreher told me. "But the truth is, it probably won't be. I know people who are so smart they make me feel like a chimpanzee. It will probably be one of them."

It could also be that the aliens will try to make life easy for us by sending out a simple mathematical message—the first ten prime numbers, for example, which should be pretty recognizable, or perhaps a series of dots and dashes that can be assembled into a graphic image of some sort. That's what Frank Drake did back in 1974 when he used the Arecibo radio telescope in Puerto Rico, which he was running at the time, as a transmitter. Drake's message included, among other things, images of the solar system, of the telescope, of a human being, and of a DNA molecule.

To make sure it was reasonably intelligible, Drake sent the message first to his friend Carl Sagan. Sagan figured it out pretty quickly, but this wasn't exactly a definitive test. Alien psychology and thinking could be so radically different from ours that it might take the average E.T. centuries to figure out what we're saying—as it might for us if the situation is ever reversed.

Without an alien to advise him, however, it was the best Drake could do. So during a ceremony rededicating the telescope after a major refurbishment, he flipped a switch and blasted it off in the direction of M-13, a dense cluster of stars in the constellation Hercules. Advertising our presence was not entirely without risk. It's not impossible to imagine—as Hollywood frequently reminds us—that aliens might be hostile, rapacious, vicious creatures just waiting for an unwary, trusting planet to announce itself. Drake's experiment was not viewed kindly by everyone who heard about it, including some scientists.

On the other hand, it will take twenty-four thousand years for the signal to reach M-13, so there's no urgency about the matter. If aliens are waiting on the other end and they have a telescope comparable in size to the one at Arecibo and they're looking at the right frequency in the right direction at the right time, then they'll see it. Assuming they choose to answer, we'll know it in about the year 50,000.

Drake has calculated that a broadcast from Arecibo could be picked up by a comparable telescope anywhere in the Milky Way. The same is true in reverse, which makes the prospect of spending time at Arecibo especially exciting for the Phoenix crew. Another SETI project, the Serendip survey already operates routinely from there. But Serendip's approach is not to lobby for its own telescope time but to attach its receivers to the telescope during unrelated observations—a sort of parasitic mode that makes it impossible for Serendip scientists to choose their targets.

Ideally, SETI researchers would love to get their hands on a telescope much more sensitive even than the one at Arecibo. They even designed such a telescope, almost thirty years ago, during a workshop held at NASA's Ames Research Center in 1971. A group of SETI-friendly scientists and engineers were given the task of deciding what it would take to mount a really serious alien detection effort. The result was an exhaustive 240-page report entitled "Project Cyclops." Cyclops was to be an array of fifteen hundred radio dishes, each of them one hundred meters in diameter—fifteen hundred football fields' worth of radio sensitivity, listening in concert for messages from the stars.

The Cyclops array never got built, largely because it would have cost $10 billion—which is $50 billion in today's inflated currency. Nevertheless, the report included the most comprehensive and tightly reasoned justification for SETI that had ever been assembled, and it bore NASA's official imprimatur. One result was that NASA formed a permanent exobiology research group, based at Ames, to address all aspects of Drake-equation science. Another effect of the report was to convince a number of young scientists that SETI was worth pursuing.

One of them was Jill Tarter, then a graduate student in radio astronomy at Berkeley. A professor named Stu Bowyer was just starting up the Serendip project, and he needed someone to program an old PDP-8S computer that he'd been given. "I was the only one around who could do it," Tarter explained. "The computer was already obsolete at that time, and believe me, the 'S' stands for 'stupid.'" Bowyer gave Tarter the Cyclops report to bring her up to speed on SETI. "I stayed up all night reading it," she remembers. "It was just fascinating to me." Eventually, Tarter moved to the Ames exobiology group, where she worked on the physics of brown dwarfs and eventually moved into NASA's in-house SETI program.

Another young scientist who was inspired by Cyclops was here at Green Bank, too, in his role as the Phoenix project manager. He wasn't down at the telescope, though. There wouldn't have been much point. Kent Cullers is totally blind and has been from birth. (Where Jodie Foster is modeled on Jill Tarter, her blind, brilliant sidekick in *Contact* is a Cullers clone.) The only computer set up to handle Braille E-mail was in his dormitory room, where he'd been holed up for days, fighting with the people down at Woodbury, Georgia, over the backup radio telescope, which is owned by Georgia Tech.

But the Georgia Tech astronomy department was evidently resentful that an outside institution had authority over its telescope. "I need to do some yelling and screaming," Cullers told me, his gentle voice belying the tough talk. "This is not my favorite thing to do."

Cullers is, as far as he knows, the only physicist in the world who was blind from birth. He was the victim of too much oxygen, a treatment routinely and, in retrospect, tragically applied to premature babies in the 1950s to help their weak lungs breathe. "I was pretty lucky, really," he told me during a lull between yelling-and-screaming calls. "I was always interested in space and astronomy. My father used to read to me from the *Golden Book of Astronomy*, among other things, and I had a natural talent for math. I could do long computations in my head. When I was in high school, in Temple City, California, they were just developing techniques for teaching science and math to the blind."

Things got a lot harder for Cullers when he moved on to Pomona College. Many university physics textbooks hadn't yet been translated into Braille. "I had to ask a million questions and corner my classmates to explain things to me. Unlike in high school, it wasn't all done for me. I also had to start understanding all sorts of diagrams—physics depends a lot on them—and I think that one extra, very difficult step is probably why there aren't any other blind physicists."

Cullers graduated with a dual degree in psychology and physics, which on their face couldn't have had less in common. Then, in graduate school,

Cullers happened on the Cyclops report. His wife read it to him, essentially without stopping. The report emphasized not only the enormous technical challenge of looking for alien signals but also the potential psychological impact—the same question that haunts Seth Shostak. "I felt," said Cullers, "that it had suddenly unified my life."

Shortly thereafter he met Jill Tarter at the wedding of mutual friends, found that she, too, was a Cyclops convert, and soon joined her at Ames. Cullers's niche in SETI turned out to be signal processing. He's the one who designed the spectrum analyzer that chops radio signals into fifty-six million channels and tests them for artificial beeps and wails. "By the year 2000 or so," he said, "this technology will be obsolete, so I'm starting to think about how we're going to upgrade it."

Cullers is also involved with a major international project that could finally create a gigantic radio array. It wouldn't come close to Cyclops's thirty-seven square miles of collecting area, but at a square kilometer it would still dwarf anything on Earth. Known as the one-KT array, it's still in the discussion stage only, with astronomers from India, China, the United States, Canada, Australia, and the Netherlands trying to figure out the best strategy for building such a device. One idea is to line a bunch of naturally bowl-shaped sinkholes in western China with radio-reflecting mesh—the same principle used at Arecibo but multiplied by a thousand. Another, suggested by the Indians, is to build one thousand more conventional antennas, using cables instead of expensive steel trusses to support the reflecting surfaces. This is referred to among radio astronomers as the "Indian Rope Trick." The Canadians, meanwhile, want to lay out an essentially flat array of reflectors on the plains and fly the receivers on a balloon three kilometers up in the air.

If the one-K ever gets built—by no means a certainty, even with the costs shared among seven countries, and potential users not limited to the SETI community—then the number of Sun-like stars within earshot is not one thousand, as it is with the 140-foot, but one million. The odds of finding intelligent, communicating life in the Milky Way will increase a thousandfold.

If the number of civilizations out there is zero, on the other hand, then you can increase your range a millionfold or a trillionfold and still find nothing. The argument that the universe is teeming with life is based only on an optimistic interpretation of the Drake equation—a compounding of probabilities that become less and less certain as you move from the right side of the equation to the left. "That's the big problem with SETI so far," says Shostak. "In most of astronomy you have all this data to argue over. In SETI you have no data of any sort." He was a little bit uncomfortable with that notion at first. He thought it would get boring pretty fast. "Actually, it hasn't at all. For one thing, you have the world's biggest carrot. Beyond that, there are always new ideas about how to look, new strategies, really ingenious stuff."

There is, however, one bit of circumstantial evidence suggesting that SETI is a complete waste of time. It was first raised in 1950 at a lunch at the Los Alamos National Laboratory, by Enrico Fermi, the Nobel Prize–winning physicist who worked on the Manhattan Project and created the nuclear reactor. Fermi's question was something like "If extraterrestrials exist, where are they?"

His point was that if the Milky Way is peppered with advanced civilizations, some much older than ours, they should long since have star-hopped into our neighborhood. Mathematically, Fermi's logic was unassailable. The Milky Way is a big place, but given enough time, it would be not only simple but almost inevitable that E.T.s should come to call. (Implicit in Fermi's question is the assumption that UFO reports have nothing to do with alien spacecraft. Essentially no serious scientist disagrees with this assertion.)

"It's really a good argument," admits Shostak. "For a while there was this whole cottage industry about the Fermi paradox. SETI people came up with all sorts of ideas on how to get out of it. There was the Drake/Oliver idea, for example: It's far too expensive to travel between the stars. It turns out you need a *lot* of energy to get from one place to another." Philip Morrison echoed this argument when we talked at Princeton.

"Then there's the 'They *are* everywhere, but we haven't noticed' idea," Shostak continued. Maybe aliens are out in the asteroid belt right now, mining it for strategic minerals. It would make sense that mineral wealth would be a whole lot more interesting to E.T.s than some wet planet. And it would be really hard to detect any activity out there. "Maybe an asteroid disappeared last night," he said. "Who would notice?"

There's the zoo hypothesis. They know we're here, and they're staying away because we're so interesting to observe and they don't want to disturb us. There's the anthropological argument: No, it doesn't take long to expand into the galaxy, but you still have to keep at it for tens of thousands of years. "If you look at Earth," said Shostak, "different groups have expanded out of their home territory, but after a while they've run out of steam. They lose interest. Their political institutions become corrupt." Maybe the same thing happens to aliens.

"Sagan argued," he went on, "that the expansion is more of a random walk than a straight line—that civilizations wouldn't necessarily expand directly outward but would zigzag and even reverse direction." That would take much longer and would imply that they shouldn't necessarily be here yet. Or maybe the galaxy is fully inhabited, but we're in some sort of no-man's-land between empires. Or maybe it's randomly populated. I mean, drive fifty miles due south of Winnemucca, Nevada, and look around. You'd swear the Earth is uninhabited.

"The problem with all of these schemes for getting around the paradox," he continued, "is that it takes only one civilization acting as an exception to the rule to ruin the whole thing. I think the Fermi paradox is still there." On the other hand, the paradox may be entirely moot for reasons that nobody has yet thought of. It's foolish to conclude that there aren't any intelligent aliens based on the Fermi paradox.

It is equally foolish to conclude that there are, based on the Drake equation. For while it's true that scientists like Marcy, Butler, Gibson, McKay, Terrile, Head, and dozens of others have finally begun to provide real data that can be plugged into an equation that was pure guesswork for thirty-five years—and into a mystery that has endured in a non-mathematical formulation for thousands of years before that—the Drake equation must soon run into an impenetrable wall.

Even if NASA's interplanetary probes find unambiguous evidence in the next decade that ALH84001 was not a fluke and that bacteria did once thrive on Mars; that extremophilic bacteria still thrive deep below the red planet's surface; that creatures swim in the dark, frigid waters of Europa's oceans; that life has arisen on three or five or seven different worlds in our solar system alone; and, finally, even if the Planet Finder proves that Earth-like planets orbiting many faraway stars are percolating with living, metabolizing, atmosphere-altering life—we still won't know for sure that intelligent life exists anywhere but here on Earth.

For it remains true, as it has since Frank Drake wrote down the equation in Green Bank, that a number close to zero anywhere in the formula could mean that humanity really is alone. And for the last two factors in the Drake equation, scientists simply have no way of coming up with any convincing number. Political scientists and historians can easily produce estimates for how long a technologically advanced civilization will last, on average, but the estimates are essentially worthless. Ours has lasted a bit less than a hundred years. It may last another million or another fifty, and any attempt to narrow that down is purely a guess. Even if we could, our single example would say nothing about civilizations in general—just as a single planet orbiting 51 Pegasi says nothing about the prevalence of Earth-like planets in the universe.

In a similar fashion, evolutionary biologists can say nothing at all about how frequently microbes would evolve into technologically advanced beings. Always—as the original conference at Green Bank had concluded? Once in a trillion times? Impossible to determine. It happened here once, but it failed to happen here many, many times. Dolphins are highly intelligent, but they haven't built cities. Primates are intelligent—they even use tools—but they haven't invented agriculture. The dinosaurs ruled the Earth for nearly two hundred million years. They were, clearly, exquisitely adapted to their environment. Some of them were probably intelligent, but they didn't develop radio. It was only the accident of an incoming comet or asteroid that cleared them out and made way for mammals to take over the planet.

Would the dinosaurs have eventually built rocket ships if the comet hadn't come or if it had been only half as large? Would the dolphins have invented telephones? Was it inevitable that humans or something like them would descend from proto-apes? The paleontologist Stephen Jay Gould says no: The sort of intelligence humans have is an accident of evolution. Technological civilization is the result of a roll of the evolutionary dice. Turn the clock back three billion years, let microbes evolve a second time, and the result would almost certainly be different. With only one roll of the dice, we can't say anything about the odds, but if Gould is right, then intelligent life could indeed be universal without revealing itself across interplanetary distances. And the only way we'll ever know that philosopher-dolphins or artist-orangutans or mathematician-dinosaurs inhabit other worlds will be to travel to the stars.

That won't be happening anytime soon, and until it does, says Seth Shostak, "you can argue back and forth until you're blue in the face about whether communicating civilizations exist. You're not going to get the answer until you do the experiment. And that's what we're going to do tomorrow—if they fix this blasted telescope."

CONCLUSION

The observing run on Mauna Kea in December 1996 has started out beautifully. At sunset a thin haze of stratospheric cirrus clouds was no more than a minor nuisance for Geoff Marcy, Paul Butler, and Steve Vogt. In order to grab the photons of light they needed, the astronomers simply held the Keck telescope on their target stars for 300 seconds each rather than the normal 120 seconds.

Now it looks as though the extra time wasn't even necessary. The charge-coupled device (CCD) detector is swamped with light. Butler steps outside, moving slowly in the oxygen-poor air, to take a look at the sky. "No wonder," he tells the others when he returns. "It's totally clear out there. No clouds at all. We're really in business now."

They are, for maybe an hour. Then Chuck Sorensen, the professional telescope operator who is working with them, notices something worrisome in a window on his computer screen and sits forward, suddenly alert. "Hey," he says with alarm, "UKIRT is saturated, you guys." UKIRT is the United Kingdom Infrared Telescope, another of the domes on the mountain. UKIRT has a humidity gauge which is wired into the computer network that links all the observatories. The gauge is reading 100 percent. The air can't hold another bit of water vapor. At best that means moisture could start condensing onto the telescope and its electronics at any second. At worst it means that it's raining or snowing outside. Sorensen grabs his coat and flashlight and runs outside in a near panic, oblivious to the danger of passing out from lack of oxygen.

Almost before the door slams he's running back in. "We've gotta close!" he shouts. The stars overhead are still shining brightly in a crystal-clear sky. Nevertheless, a fine sleet is pelting the Keck building, blown up to this altitude from the solid overcast hovering a couple of thousand feet below the summit. The telescopes are in imminent danger of icing up, like the wings of airplanes sitting on a subfreezing runway. Sorensen quickly taps in a series of commands that will shut the dome tight against the elements. The astronomers keep taking data until the very last second, pulling in photons even as the doors slam shut.

The dome will remain shut for three hours, although for all anybody knows the sleet could last all night. Observers at the other telescopes phone in to say they're bailing out, heading downhill to Hale Pohaku and a game of pool or a snack or some sleep. "What a bunch of wimps," Butler says with a sneer. The others agree. They've come all the way to Hawaii and aren't leaving if there's even a remote chance of getting some more observing in. And the sky overhead is, after all, tantalizingly clear.

Besides, telescope time has become more precious than ever in the last few months. As long as they were looking for planets in vain, laboring in nearly total obscurity on a project that seemed to be going nowhere, Marcy and Butler didn't have to worry much about the competition. Most scientists thought finding planets was simply too hard, and the few who didn't were years behind. Michel Mayor managed to sneak in ahead of them, but that was only because 51 Peg B was such a bizarre object.

Once Marcy and Butler actually started finding things, however, planet hunting became the hottest thing in astronomy, essentially overnight. Everyone wanted to do it. "A whole bunch of people are suddenly saying, 'This is easy,'" says Butler. "They're all trying to get into it. They'll find out it isn't easy at all." But just as happened when Roger Bannister ran the world's first four-minute mile, the proof that a goal is attainable is a powerful incentive to try it yourself. And despite Butler's dismissive attitude, a lot of his competitors, especially those who have been in the planet-hunting game for a while, will probably succeed.

Thus, while he knows Mike Shao will be kicking everyone's butt in a few years, Paul Butler wants to make sure he and Marcy and Vogt dominate the field until then. In fact, during a conversation with Marcy a few months earlier he'd argued for an observing strategy that would guarantee it. He'd wanted to add another sixty stars to the Lick survey right away and observe them with the same level of precision they'd used to find their first six planets. "I say we clean up all the easy guys, the plums," he'd said. "On a one- or two-year time scale I don't see anyone touching us." They would corner the market in big, easy-to-find planets.

"I don't know," Marcy had objected. "I think plums are starting to get boring. They're out there, we know they're out there, we know how frequently they occur." He wanted to put all their intellectual energy into the Keck observing program instead, into mastering the giant telescope and the new spectrometer in order to find the Jupiters and Saturns and Neptunes and Uranuses that nobody else could. Let the others console themselves with finding 51 Peg Bs and the like; the world's greatest planet-hunters would take the game to a new level. They'd go for the three-minute mile.

Marcy eventually won the argument, which was probably moot in any case. There was no way he and Paul Butler could observe all or even most of the promising stars in the sky. Other talented observers would find plenty of planets if they were out there.

In fact, several already had. A few months before the Keck run, Bill Cochran, a planet-hunter (and Origins committee member) from the University of Texas, had identified a planet orbiting the star 61 Cygni. It was only a little bigger than Jupiter, with an average orbit a little farther out than Mars's, but it had the most eccentric orbit yet seen. The theorists were still trying to fit the object into their ideas about how such an object could exist.

Marcy and Butler had found the planet, too, but gave Cochran the honor of announcing it and attaching his name to the discovery. He was a friend, a good guy in their book—they'd even helped him build his own iodine cell so he could copy their technique.

A few months before that—at the June meeting of the American Astronomical Society, which followed Marcy and Butler's triumphant appearance at the group's January conference—George Gatewood announced his own discovery. He'd found at least one, and maybe two, planets orbiting the star Lalonde 21185—one of the stars displayed on a plastic model that sits in his office. He was terrified to announce it since he made his reputation by shooting down others' planet discoveries. Worse yet, he'd debunked a planet around this very same star, back in the 1970s. This time David Black, who had always advised Gatewood to keep quiet about possible planet detections in the past, had advised him to go ahead.

Gatewood found his planet with Allegheny's old-fashioned thirty-inch refractor, a near-twin to the thirty-six-inch at Lick, and used old-fashioned astrometry, measuring side-to-side movements, not radial wobbles, to do it. He uses an idea that Frank Drake once suggested: Gatewood lets starlight shine through a series of fine slits before falling on the CCD, and then lets the stars drift past the telescope. As they do, they wink in and out of view. The pattern of winking gives Gatewood a precise measurement of where on the sky each star is and whether it's wobbling in place. So far he is the only astronomer using this technique since, like Marcy and Butler, he's the only one who has built the right sort of detector. Now that he has shown it works, others are interested in trying it.

Harvard's David Latham is also counted among the ranks of planet-finders, not by his own assertion but by that of Geoff Marcy and Paul Butler. The object Latham and then Michel Mayor identified in 1989, the "low-mass object" HD114762, seems, on closer examination, to be a big planet, not a brown dwarf. So Marcy and Butler believe, at least, and so they say on their Web site. Like Cochran and Gatewood, Latham is following Marcy and Butler to the Keck so he can start observing stars wholesale. (And like Cochran, who sits with him on the Keck Telescope Allocation Committee, he recuses himself from decisions on his own application for telescope time.)

If Latham's application is approved, he, like all the other radial-velocity searchers, will be using Steve Vogt's high-resolution spectrometer. The instrument belongs to Vogt only in the sense that a grown child belongs to a parent. He brought the high-res into the world, but he has no control over it.

The same goes for Marcy and Butler's Keck iodine cell. Shortly after the confirmation of 51 Peg, the people at Keck contacted the two astronomers to ask them, as the inventors of the iodine cell, to make one for the Keck. Marcy agreed. Given the competitive nature of planet hunting, I found that slightly odd when I first heard about it. During the bioastronomy conference on Capri, I finally cornered Marcy and asked him how this was going to work.

He was something of a captive audience. The conference organizers had put together a boat tour around the island. Halfway around, the boats stopped and anchored beneath Capri's towering cliffs. The bathing-suited scientists began diving off the bows into the cool, perfectly blue Mediterranean. As we sat on deck afterward drying off, I quizzed Marcy about the iodine cell. Would he really spend his precious time building an instrument that would be used mostly by his competitors? Who would actually own the thing? Would he have a veto over who could use it, and if so, wouldn't that be a conflict of interest?

"Well," he said, a thoughtful expression crossing his face, "the truth is that I hadn't really figured all that out yet. They asked me to build it, so I said yes. But I really have to work out those administrative details, and, frankly, I'm not sure how these things work." A woman sitting on a nearby chair spoke up. "Geoff, there are established guidelines for this sort of thing," she said. It was Jill Tarter, who was on Capri to give a talk about Project Phoenix. "What you want to do," she said, "is work out an arrangement where they pay you a fair price for your time and your expertise. Then the instrument belongs to them, and anybody who wants it gets to use it." Marcy mulled that over and decided it was reasonable. This is exactly the arrangement he worked out with the Keck administration. The observatory paid him—his research group, really—about $200,000 for the iodine cell, and it's now a Keck instrument.

As a European, Michel Mayor is one of the few planet-hunters who can't apply directly for NASA time on the Keck. He may join Dave Latham's consortium, but lack of access to the world's most powerful telescope has not held him back. During the bioastronomy conference on Capri, Mayor had stood up and announced the discovery of eight new objects, all of them at least twenty times as massive as Jupiter. These were clearly brown dwarfs, not planets, but they lay right in the so-called desert that Geoff Marcy had claimed distinctly separated true brown dwarfs from true planets. If the desert wasn't empty, then maybe 70 Vir and the other giant planets were indeed just the smallest of the brown dwarfs, as Dave Black was still insisting.

"No, no, no," Marcy almost shouted when I asked him about it. "We never said there aren't *any* objects in the desert, just that there are many fewer than you'd expect. Yes, Michel has found eight objects, compared with the seven or eight new planets. But—and this is really an important point— *those are from a much bigger survey.*" If you correct for the survey size, there are still too few medium-size brown dwarfs. Besides, Mayor's discoveries come from low-precision surveys; there might well be dozens of 70 Virs lurking there that he never would have picked up. "Michel's objects are very important, but they have no bearing on our assertion that we've been finding true planets."

Another set of observations, however, could have a very serious bearing on Marcy and Butler's discoveries, and on Mayor's and Queloz's original detection of 51 Peg B as well. Toward the end of February 1997, eighteen months after the 51 Peg announcement and a bit more than a year after the announcements of 70 Vir B and 47 UMa B, a report appeared in *Nature* which said, in essence, that Mayor, Queloz, Marcy, and Butler were wrong.

The accuser was David Gray of the University of Western Ontario, an expert on stars. "I've been monitoring 51 Peg for years," he told me when I spoke to him on the phone just before the *Nature* paper came out. "The changes those guys see in the star's spectral lines don't come from a planet. They come from the star itself. 51 Peg is variable. The planet isn't there."

But how was that possible? The proposition that 51 Peg was simply expanding and contracting, the in-and-out motion of its surface masquerading as a forward-and-back wobble of the entire star, was the most obvious objection, and Mayor, Queloz, Marcy, and Butler had addressed it easily. 51 Peg shines with a steady brightness. Obviously, it wasn't pulsing.

Nevertheless, insisted Gray, it was a variable star. His evidence lay in 51 Peg's spectral lines, the markers that planet-finders examine to tease out a star's radial velocity. If you look at the lines very carefully, said Gray, you'll see that they not only shift positions, first redward and then blueward, but actually change shape as well. The lines aren't perfectly straight to begin with, which is normal. But these lines wiggle. They change their tilt. They bend and unbend. "There's simply no way a planet could do that," says Gray. "The planet hypothesis is slain. No question about it."

Yet 51 Peg is demonstrably not changing its brightness, so if it's varying, it must be varying in a way no astronomer has ever seen before. Gray's suggestion is that the star's surface is moving not in and out but back and forth, like a loose skin. And he has computer simulations to show that if this were happening, it would produce the effects he sees in the real star's spectral lines.

The suggestion that their first planet doesn't really exist—and, by implication, that at least some of the others don't, either—had Marcy and Butler incensed. Even before Gray's paper formally appeared, their Web site was carrying a detailed refutation. They pointed out, among other things, that real stellar pulsations almost always come in sets—the star vibrates at several different frequencies at once, just as a violin string emits overtones in addition to its primary pitch. 51 Peg's variations come in just one frequency: four days. Gray, they conclude, is just plain wrong.

In fact, most astronomers who have looked at the evidence think that both extremes—no planet, on the one hand, and no point in listening in Gray, on the other—are premature. Gray has enough standing as a variable-star expert that his assertion is worth following up with new observations, but he hasn't come close to showing that 51 Peg B is all a mistake.

Even if Gray turns out to be right, his arguments are unlikely to apply to stars like 70 Vir and 47 UMa, where any "pulsations" would have to be unbelievably slow to account for the planets' long orbital periods.

So while scientific discoveries always leave some small room for doubt—physicists still test even general relativity from time to time on the off chance that Einstein slipped up in some subtle way—it's a pretty safe bet that other worlds are out there. The new worlds are, we presume, too big or too hot to sustain life as we know it—even the sort of life tough enough to survive under the surface of Mars or deep in Europa's oceans. But our knowledge of other worlds has just begun to accumulate. Over the next few years it will be increasing at an exponential rate as more astronomers get involved in the search, as new, more powerful telescopes and spectrometers and CCDs are built, as NASA's Origins Program moves from the Earth into space and per-haps all the way out to Jupiter's neighborhood.

Giant space telescopes and sensitive interplanetary interferometers, moreover, are only the most conventional means astronomers will use to search for and probe the characteristics of other worlds. Several groups pro-pose to look not for a star's wobbles but for the nearly imperceptible dim-ming that a planet would cause as it passed in front of its star. Others are already looking for stars that show a particular sort of twinkle—caused by the slight bending of starlight by an orbiting world's gravity, which can turn a planet briefly into a sort of magnifying lens.

A dozen other equally unconventional notions will undoubtedly surface before long. Astronomers, as Geoff Marcy and Paul Butler have proven, are never at a loss for new ways of attacking an old problem. Paul Butler sug-gested soon after he and Marcy had found 70 Virginis B that there would soon be more planets known outside the solar system than inside. He was right. Now he's predicting that we'll have evidence for hundreds of extrasolar planets by the year 2000. He may be too conservative.

What Butler refuses to predict is where he'll be by then. Marcy will be staying at San Francisco State for the foreseeable future. A couple of weeks after all the excitement happened, San Francisco State gave Marcy a big raise. "I believe," says Butler, "they've retroactively boosted his salary by more than I'm being paid.

"Nobody is rushing to hire me, though. I'm still only a postdoc, which is a finite position. The only job I want in the entire world right now is a senior observer position at the University of Hawaii, and three hundred other peo-ple want it, too. Technically, I don't meet their minimum requirements, but out of pure bravado I tossed off an application anyway. I'll see what happens. If I'm unable to get a job, the hell with it. I'll have had my little run, you know? I remember when I got out of school, I was really worried about the statistics. But I don't worry about that anymore. If things pan out, great, and if not, so what?

"Anyway, when people in astronomy start working outside astronomy, they typically double their income. But I don't like working nine to five or having a boss. I don't have great people skills. Science provides a place for an antisocial person like me. It gives me a place to hang out and do what I want and contribute to society. And in return, society gives me a salary right up there at the level of a senior janitor."

Then, as we wait for the overcast to clear, Paul Butler tells me about this crazy new idea he has. Until the Planet Finder goes into orbit, nobody has much hope of seeing any planets directly—only indirectly, through their gravitational tugging on stars. But he, Geoff Marcy, Steve Vogt, and Bob Noyes at Harvard (Butler is temporarily suspending his contempt for that institution) think they can do it from the ground with an ordinary telescope.

The idea is that 51 Peg B and the other tightly orbiting planets are so close to their stars that they ought to reflect a lot of light—almost as though they were tiny mirrors. Whatever you see in the star's light, spectral lines included, you should also see bouncing off the planet. But while the star is wobbling back and forth at a few tens of meters per second, the planet is whipping around the star at many thousands of meters per second. That means its reflected version of 51 Peg's light, while puny in intensity compared with the star's, should be Doppler shifted much more drastically.

The two signals should be distinguishable, in principle, and once you disentangle the star's light from the reflected light, says Butler, you can find out all sorts of things about the planet—its size (by how bright the reflected light is), its composition (by how reflection alters the original starlight), its density, and all sorts of other things. "It'll be a really, really hard measurement," says Butler with a grin. "It's probably impossible, in fact. I can't wait to get started."

A NOTE
ON SOURCES

This is a work of journalism rather than a work of scholarship. Its purpose is to give readers a sense of what's going on right now in one of the most exciting areas of current science. As such, it is based largely on interviews with its leading characters, the astronomers and other scientists who are searching for and thinking about life on other worlds. These took place, for the most part, in the places where the searching and thinking and designing of new instruments is being done—at Lick Observatory, Palomar Observatory, the Jet Propulsion Laboratory, the SETI Institute, NASA Ames Research Center in California; at the Keck Observatory in Hawaii; at the Steward Observatory Mirror Lab in Arizona; at the Goddard Space Flight Center and the Space Telescope Science Institute in Maryland; at Johnson Space Center and the Lunar and Planetary Laboratory in Texas; at NASA headquarters and the Carnegie Institution in Washington, D.C.; at Allegheny Observatory in Pennsylvania; at the National Radio Astronomy Observatory in West Virginia; at universities around the country and conferences around the world.

Unlike a scholarly work, however, it is not laden with references to scientific papers, historical documents, or private papers, nor does it make any attempt to be an exhaustive review of the topic that touches every conceivable base. For those who are interested in looking more deeply into some of the questions and issues raised in this book, I offer a partial list of the references I found most useful.

Angel, J. Roger P., and Neville J. Woolf. "Searching for Life on Other Planets," *Scientific American*, April 1996.

Beichman, C. A., ed. *A Road Map for the Exploration of Neighboring Planetary Systems* (JPL Publication 96-22). Houston, Tex.: Jet Propulsion Laboratory, 1996.

Bowyer, Stuart, Cristiano Cosmovici, and Dan Wertheimer, eds. *Astronomical and Biochemical Origins and the Search for Life in the Universe: Proceedings of the Fifth International Conference on Bioastronomy.* Bologna, Italy: Editrice Compositori, 1997.

Davies, Paul. *Are We Alone?: Philosophical Implications of the Discovery of Extraterrestrial Life.* New York: Basic Books, 1995.

De Duve, Christian. *Vital Dust: The Origin and Evolution of Life on Earth.* New York: Basic Books, 1995.

Dick, Steven J. *The Biological Universe: The Twentieth-Century Extraterrestrial Life Debate and the Limits of Science.* Cambridge, Mass.: Cambridge University Press, 1996.

Drake, Frank, and Dava Sobel. *Is Anyone Out There?: The Scientific Search for Extraterrestrial Intelligence.* New York: Delacorte Press, 1992.

Dressler, Alan, et al. *HST [Hubble Space Telescope] and Beyond.* Association of Universities for Research in Astronomy, Washington, D.C., 1995.

North, John. *The Norton History of Astronomy and Cosmology.* New York: W. W. Norton, 1995.

Sullivan, Walter. *We Are Not Alone: The Search for Intelligent Life on Other Worlds.* New York: McGraw-Hill, 1964.

Swift, David W. *SETI Pioneers: Scientists Talk About Their Search for Extraterrestrial Intelligence.* Tucson: University of Arizona Press, 1990.

PHOTO CREDITS

Goddard Space Flight Center, NASA: 11–13; Jeff Hester and Paul Scowen (Arizona State University), NASA: 15; Jet Propulsion Lab, NASA: 7–9, 14, 18; Michael D. Lemonick: 1, 2, 5, 16, 20; Mark McCaughrean (Max Planck Institute for Astronomy), C. Robert O'Dell (Rice University), and NASA: 3; Andrew Perala: 6; Seth Shostak, SETI Institute: 19; Space Telescope Science Institute: 17; Steward Observatory, University of Arizona: 10.

INDEX